Essential Atlas
of
Physics and Chemistry

BARRON'S

First edition for the United States, its territories and dependencies, and Canada published 2004 by
Barron's Educational Series, Inc.
© Copyright of English-language translation 2004 by Barron's Educational Series, Inc.

Original title of the book in Spanish: *Atlas Básico de Física y Química*
© Copyright 2003 by Parramón Ediciones, S.A.—World Rights
Published by Parramón Ediciones, S.A., Barcelona, Spain

Author: Parramón's Editorial Team
Illustrator: Parramón's Editorial Team
Text: Jordi Llansana

English translation by Eric A. Bye

All inquiries should be addressed to:
Barron's Educational Series, Inc.
250 Wireless Boulevard
Hauppauge, NY 11788
www.barronseduc.com

ISBN-13: 978-0-7641-2713-7
ISBN-10: 0-7641-2713-6

Library of Congress Catalog Card No. 2003114846

Printed in Spain
9 8 7 6 5 4 3

PREFACE

The purpose of this atlas is to give students a basic familiarity with physics and chemistry but without emphasizing the mathematical content and formulas any more than necessary. The intention is to explain the concepts clearly and simply since these are the major stumbling blocks for young students.

The editing of this book has avoided an excess of data that would not contribute to the reader's understanding or provide an idea of any magnitude. However, some interesting facts have been emphasized because they are of extremely little or extremely great value. We have also attempted to establish the relationship, to the extent possible without distracting the reader from the main idea, between the theoretical world of physics and chemistry and daily life. We are convinced that these two sciences are not confined to laboratories but, rather, are part of our daily lives.

We will be gratified if students discover this atlas to be of real help as an important complement to other texts, which it is not intended to replace, and as a guide for getting to know and love nature by learning to respect it and preserve it.

TABLE OF CONTENTS

Introduction 6

Forces and Their Effects 10
What Causes a Force?............................... 10
   Force Defined.. 10
   Hooke's Law 10
   The Vector Nature of Forces 11
   Representing a Force 11
   Measure and Moment of Forces 11
How Do Several Forces Work Together? 12
   The Additive Nature of Forces................ 12
   Forces Applied at a Single Point 12
   Parallel Forces 13
   The Decomposition of a Force 13
Forces at a Distance.................................. 14
   The Law of Universal Gravitation 14
   Gravity 14
   Weight 14
   Magnetic, Electrostatic, Electromagnetic,
   and Nuclear Forces............................... 15
Simple Machines.. 16
   The Lever 16
   Pulleys and Winches............................. 17
   The Inclined Plane and the Screw........... 17

Motion.. 18
Reference Systems 18
   When Does Motion Occur? 18
   The Relativity of Motion 18
   Reference Point 19
   Reference System 19
   The Principle of Relativity 19
Elements That Characterize Motion 20
   Kinematics, Position, Trajectory, and Displacement . 20
   Mean Velocity and Acceleration 21
Types of Motion 22
   Trajectory and Velocity 22
   Rectilinear Motion.................................. 23
   The Motion of Falling Bodies..................23
   Vibratory Motion 23
   Uniform Circular Motion 24
   Acceleration 24
   Pendular Motion 25
   Periodic Motions 25
The Composition of Motions 26
   Horizontal Motion and Oblique Motion 26
   Perpendicular Vibratory Motions 26
   Same Frequency and Maximum Expansion 27
   Same Frequency and Different Expansion 27
   Different Frequency 27
Force and Motion 28
   Newton's Principles 28
   The Acceleration of Gravity 28
   The Force of Friction.............................. 29
   Rope Tension 29

Energy 30
Work, Power, and Energy 30
   Energy and Types (Fossil, Renewable)..... 31
Mechanical Energy 32
   Gravitational Potential Energy 32
   Kinetic Energy 32
   Quantity of Motion 33
   Elastic and Inelastic Impacts 33
The Conservation of Energy 34
   The Conservation of Mechanical Energy 34
   Calculating Velocities and Heights 34
   Work as an Energy Modifier 34
   Is Energy Really Invariable? 35
Heat, a Source of Energy............................ 36
   Transforming Work into Heat and
   Heat into Work 36
   Thermal Machines 36
   Steam Engines 37
   Internal Combustion Engines 37
The Effects of Heat 38
   Expansion 38
   The Heating of Objects 39
   Heat Absorbed by an Object 39

Fluids 40
Characteristics of Liquids 40
   Forces of Cohesion 40
   Liquids and Containers 40
   Surface Tension 41
   Capillarity and Viscosity 41
The Theorems of Pascal and Archimedes 42
   Density and Pressure............................. 42
   Communicating Vessels 42
   Flotation 43
Gases 44
   The Pressure-Volume Relationship 44
   Absolute Temperatures 44
   Ideal Gas Law....................................... 45
   Atmospheric Pressure and Its Effects..... 45

Wave Motion 46
Characteristics of Waves 46
   Types of Waves 46
   Wavelength 47
Sound 48
   Properties of Sound 49
   Intensity of Sound 49
   Tone and Timbre 49
Optics 50
   The Nature of Light................................ 50
   How Does Light Move?........................... 50
   Shadow and Half-light (penumbra) 51
Mirrors 52
   Types and Laws of Reflection 52
   Types of Mirrors 52

Lenses 54
   Refraction.. 54
   Types of Lenses 54
   Convergent and Divergent Lenses 55

Electricity **56**
Electrostatic Phenomena 56
   Types of Charges 56
   Coulomb's Law.................................... 56
   Intensity of Field 57
   Electrical Potential 57
Electrical Current...................................... 58
   What Creates the Difference in Potential? 58
   Types of Current 58
   Electrical Magnitudes 59
   The Effects of Current 59
Magnetism and Electromagnetism 60
   Magnets 60
   Magnetic Induction 60
   Current and Magnetism 61
   The Law of the Right Hand 61

Matter **62**
The Classification of Matter 62
   Heterogeneous Mixtures...................... 62
   Homogeneous Materials 63
   Pure Substances 63
   Symbols and Formulas 63
Kinetic Theory 64
   Solids 64
   Liquids and Gases 65
   Laws and Motion 65
   What Is a Change of State? 66
   Vapor Pressure 66
   Fusion 66
   Boiling 67
   Evaporation and Boiling 67

Inside Matter **68**
Atoms 68
   Gases 68
   What Are Atoms Like? 69
   The Mole 69
Periodic Classification 70
   The Periodic Table 70
   Atomic Radius 70
   Ionization Energy 70
   Electroaffinity.................................... 70
   Oxidation Number 71
Bonds 72
   Chemical Bonds and Ionic Bonds 72
   Types of Covalent Bonds..................... 73
   Metallic Bonds 73
Giant Structures 74
   Ionic Solids and Covalent Solids 74

   Diamonds.. 74
   Graphite.. 75
   Polymers 75
Mixtures 76
   Types of Solutions 76
   Variation in Solubility 77
   Ways of Expressing Concentration 78
   Suspensions and Gels 78
   Emulsions.. 79
   Foams and Aerosols 79

Pure Substances **80**
Inorganic Substances 80
   Oxides 80
   Hydroxides 81
   Acids 81
   Salts 81
Carbon Compounds 82
   Carbon Chains 82
   Hydrocarbons 82
   Compounds with Oxygen 83
   Nitrogen Compounds 83
The Chemistry of Life 84
   Carbohydrates 84
   Polysaccharides 84
   Lipids and Proteins 85

Chemical Changes **86**
The Most Important Chemical Reactions 86
   Mixtures and Chemical Reactions.......... 86
   Acid-Base Reactions 87
   Oxidation-Reduction Reactions 87
   Combustion Reaction 87
   Equilibrium Reactions 87
Energy in Chemical Reactions 88
   Why Does Reaction Heat Exist? 88
   Expansion Work 89
   Electrical Energy................................. 89
   Electrolysis 89
Chemical Industries 90
   Industry 90
   Cement 90
   Sodium Hydroxide 90
   Ammonia and Polymers....................... 91
Secrets for Discovery 92
   Nuclear Fusion 92
   Biogas 92
   Magnetic Heating 92
   Artificial Intelligence 93
   Superconductors 93
   Superfluids.. 93
   Quantum Computers 93
   Holographic Images............................ 93

Alphabetical Subject Index.................. **94**

TABLE OF CONTENTS

INTRODUCTION

THE FIRST SCIENTISTS

It is impossible to specify when physics and chemistry began, but we can state that they originally did not exist as two separate sciences. No scientists existed. There were only people with their ability to reason and a few necessities, such as eating, clothing themselves, curing sicknesses, and so forth, which they had to master as quickly as possible to improve their quality of life. At first, people **used nature** to feed themselves and make their tools: animal bones and stones were adapted to the hunt. Surely in thunderstorms they saw that nature was capable of causing **fire** with its thunderbolts. Little by little, and undoubtedly with many false starts, they reached their goal and succeeded in reproducing this wonderful fire. From that moment on, meat, the product of their hunting efforts, would not go bad as quickly. Consider that these primitive people were the **first chemists, physicists,** and **biologists**. They discovered that you can produce heat by rubbing two pieces of dry wood together. With combustion, they caused an artificial chemical reaction, and with fire they found a way to eliminate the bacteria that produce putrefaction and the accompanying foul smell.

The discovery of copper and iron marked evolutionary steps for primitive humans.

The discovery of fire was the first stone in the great edifice of science.

Nature also showed people the existence of **metals**, which revolutionized their lives. At the beginning, people recognized the scarce metals that are found free in nature, such as copper and gold. With the aid of fire, they learned to produce copper from certain bluish rocks that leave a metallic residue when they are heated. This was the first reaction of **metallurgy**. With the advent of metals, people studied their properties and found practical applications in the home, the field, and arms. Soon they discovered other metals and learned to mix them and produce the first **alloys**. Mixing copper with tin produced bronze around the third millennium B.C. This metal, which is harder and stronger than copper, was an ideal raw material for arms and shields. Iron was already known through the remains of meteorites, but it was so scarce that it was not usable. Iron was more tightly bonded to the rocks that bore it than was copper. However, one day someone had the idea of burning these rocks with wood charcoal and pro-

Introduction

Forces and
Their Effects

Motion

Energy

Heat

Fluids

Wave
Motion

Sound

Optics

Electricity

Matter

Inside
Matter

Mixtures

Pure
Substances

Chemical
Changes

Alphabetical
Subject Index

For the ancient Greeks, water was the most important element for no life can exist without it.

duced pure **iron** that was harder and stronger than copper. They even discovered that if carbon was added to the iron, the result was a harder and less fragile metal than pure iron (cast iron); that is how **steel** was introduced.

THE GREEKS

The Greeks were not great explorers, but they certainly were great thinkers about the nature of matter. In the sixth century B.C., Thales, a philosopher who was born in Miletus, observed that substances could be transformed. He thought that all substances were the same things that could take on different appearances. Thales called this basic material the **element**. Thales believed that the earth was a flat disk covered by the celestial hemisphere and that it floated on an infinite sea. Since water was the most important thing, not only because it made up the ocean, the rivers, and the atmosphere but also because life cannot exist without it, Thales said that the element was **water.** His hypothesis had many followers in his time and in following centuries. A little later, Anaximenes, also from Miletus, believed that the basic element or substance was **air**, which surrounds everything in existence. When it was compressed, it formed water, and if it was compressed further, it formed dirt. Heraclitus, who was born in Ephesus, near Miletus, believed that the element was that

which experienced the greatest changes and caused the greatest changes: **fire**. Almost a century later, Empedocles, who was born in Sicily and was an outstanding disciple of the mathematician, physicist, astronomer, and philosopher Pythagoras, suggested that the elemental theory was too simplified and that nature was made up of **four elements**: the water of Thales, the air of Anaximenes, the fire of Heraclitus, plus his addition, **earth**. Even Aristotle, a full century

Although the Greeks believed that the atom was an indivisible particle, in the first half of the twentieth century humans succeeded in splitting the nuclei of atoms to produce tremendous energy (in nuclear power plants for electricity production) and highly destructive weapons (the atomic bomb).

7

INTRODUCTION

Alchemists combined the theory of the four elements with the heavenly bodies that represented various metals.

later, accepted the theory of the four elements. However, he presumed that each of them was made up of a **pair of properties** from among the four following opposites: cold, heat, wet, and dry. These can be used to form four different pairs that are the various elements: heat and dry are fire; heat and wet are air; cold and dry are earth; and cold and wet are water.

In the same era, there arose another topic for discussion that strikes us as being much more modern. This discussion centered around what would happen if a piece of matter were cut in half and one of these halves were cut in two and so forth. Naturally, there were two conflicting opinions. For Leucippus and his disciple Democritus, the end result would be a piece so tiny that it would be indivisible; this they called an **atom.** For many other philosophers, including Aristotle, matter was divisible ad infinitum.

ALCHEMY

In the fourth century B.C., the theories of the Greeks blended with Egyptian scientific practices. It could have been a fruitful combination, but Egyptian science was strongly influenced by **superstition**, so it remained in the hands of wizards or sorcerers who created a dark world that was partly feared and partly admired. This obscurity in the scientific world hindered progress and gave rise to charlatans and frauds. Around 200 B.C., the first alchemist was Bolos of Mendes, who began studies of **transmutation**. In other words, he sought a formula that would make it possible to transform metals into gold. This was the driving force of alchemists until after the Middle Ages. During those years, scientists made important discoveries. However, they were unconnected to one another, and advancement was very difficult and incomplete.

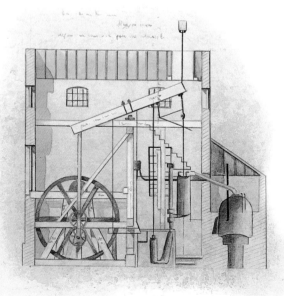

The application of a certain law of physics (the expansive force of water converted to steam by heating) made it possible to create the steam engine, which was essential to the progress of technology.

Forces and
Their Effects

Motion

Energy

Heat

Fluids

Wave
Motion

Sound

Optics

Electricity

Matter

Inside
Matter

Mixtures

Pure
Substances

Chemical
Changes

Alphabetical
Subject Index

PRESENT-DAY SCIENCE

One invention was destined to change the direction of human knowledge: Johannes Gutenberg's printing press in the fifteenth century. This contrivance made it possible to publish books, so discoveries could be shared with other scholars. With Galileo at the end of the sixteenth century, Newton in the seventeenth, and Lavoisier and Boyle, measurement became important. A long, qualitative stage in scientific knowledge drew to a close as the foundations for modern science were established. In the last five centuries, science and technology have made much greater strides than in all the dozens of preceding centuries. Wise men exited the stage, and nowadays knowledge is much more specific. There are chemists, physicists, biologists, and others. Within each of the sciences are specialized subdivisions. Chemistry may be inorganic or organic, depending on whether the object of study is from the mineral world or the realm of living creatures. Each specialty has successive subdivisions; thus an

A large portion of the electricity that reaches industry and our homes comes from thermal power plants. They release the energy contained in carbon compounds, such as coal and oil.

organic chemist may be a specialist in biochemical procedures, petrochemicals (products made from petroleum), polymers, pharmaceutical products, and so forth. The situation in chemistry also exists in physics. It too has become subdivided into solid mechanics, fluid mechanics, wave mechanics, quantum mechanics, and others. Knowledge has become so broad that it is unlikely or even impossible that any one mind could contain it all and make use of it. Modern scientists have a slight knowledge of all similar sciences and one specialty that they know and study in depth. Their work is not an end unto itself but, rather, it complements and serves the work of other scientists.

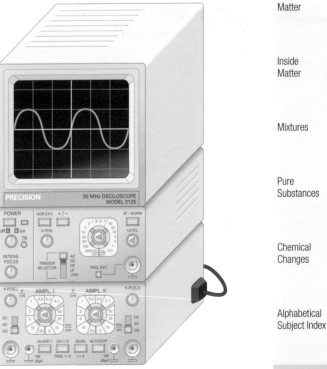

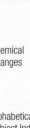

Electronics is the branch of physics that studies and uses the variations in electrical magnitudes (such as currents, charges, electric tension, and electromagnetic fields) to capture, transmit, and take advantage of information. The illustration shows an oscilloscope.

WHAT CAUSES A FORCE?

Force is a quantity present in all parts of the universe. Phenomena as different as the movement of the planets around their star, the formation of mountains, the tides, falling snow, the movement of cars, and the speed of the electric charges in lightning are all due to forces. These common occurrences are not the result of a single force but of a common and simultaneous action involving several forces with a single outcome.

FORCE DEFINED

Forces can be defined and studied only through their effects. A force is everything that is capable of changing the state of **rest** or **motion** of a body. Thus, motion can be imparted to a body at rest by one or more forces. Additionally, a body in motion can be stopped, as if by applying the brakes, accelerated, or turned off course. A force can also **distort plastic** or **elastic** bodies, and finally, it can also alter the effects of another force.

The formation of a mountain or a lightning bolt requires the action of multiple forces.

HOOKE'S LAW

If a spring is deformed, whether through **extension** or **contraction**, it will exert a recovery force that returns it to its original shape. This force is directly proportional to the deformation produced. Hooke's law is expressed mathematically as $F = -K\Delta l$.

The minus sign in Hooke's law is because the force exerted by a compressed or extended spring always takes a direction opposite that of the deformation.

In order to stretch the exercise cable, the athlete has to exert a lateral force proportional to the desired stretching.

The K, or Hooke's constant, of a spring depends on the material from which the spring is made, the thickness of the wire, the diameter of the coils, the distance between coils, and the temperature.

The tremendous force of the motors will soon place the spaceship at a high altitude.

Forces and
Their Effects

Motion

Energy

Heat

Fluids

Wave
Motion

Sound

Optics

Electricity

Matter

Inside
Matter

Mixtures

Pure
Substances

Chemical
Changes

Alphabetical
Subject Index

UNITS OF FORCE

SYSTEM	UNIT	SYMBOL	EQUIVALENCIES
International (SI)	Newton	N	1 kp = 9.816 N
Centimeter-Gram-Second (CGS)	dyne	dyn	1 N = 100,000 dyn
Technical (TS)	kilopound or kilo force	kp	

THE VECTOR NATURE OF FORCES

Forces are **vector quantities** because in order to determine their effects, it is necessary to indicate their **intensity**, where and in what **direction** the force is applied, and **what line of motion the object follows**.

Intensity

Line of Motion

Point of Application

Direction

REPRESENTING A FORCE

A force is represented by a vector. A vector is an arrow whose length represents the **intensity** (a number and a unit). The origin of the arrow represents the **point of application** of the force, the straight line of the arrow indicates the **line of motion**, and the point shows the **direction**.

MEASURING FORCES

Forces are measured using a **dynamometer**. This instrument compares a force with the lengthening effected in a calibrated spring.

Calibrated Spring

Pointer

Scale in N

Measuring a force with a dynamometer is an **indirect measurement** since we are not comparing magnitudes of the same type. We learn about a force by means of length.

6N

A magnitude is scalar when it can be specified using a **number** and a **unit**. For example, your age can be specified using a number and the word *years* for the unit.

THE MOMENT OF A FORCE

The moment of a force with respect to a point is the product of the intensity of the force multiplied by its distance from the point. The moment of a force measures the capacity of the force to produce a turn.

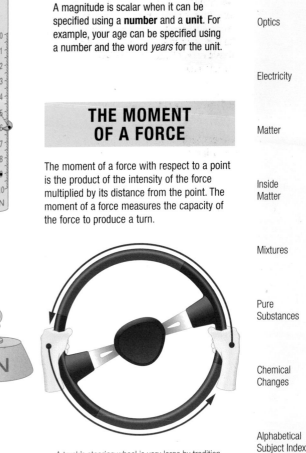

A truck's steering wheel is very large by tradition. In earlier times, the diameter was very large so a person's arms could turn the wheel.

The refrigerator handle is located at the edge opposite the hinges so that the force will be greater.

HOW DO SEVERAL FORCES WORK TOGETHER?

Rarely do we encounter a real phenomenon that is the product of a single force. Consider that we live in a world in which there are two practically inescapable forces: **weight** and **friction**. Thus, it is useful to be able to predict the effects that various forces will produce when acting on an object simultaneously and to know how to determine the forces that cause a certain effect.

THE ADDITIVE NATURE OF FORCES

The resultant or **sum of forces** is a single force that produces the same phenomenon as several forces acting simultaneously. The method used in determining the sum depends on the situation of the forces: whether they are applied to one point or to a **rigid solid object**, that is, whether they converge at a single point or act in parallel.

The three men together produce the same effect as the mule.

FORCES APPLIED AT A SINGLE POINT

Two forces applied to the same point can be combined using the **parallelogram rule**: the two forces are represented using their respective arrows. From the end of one of the lines of force a line is drawn parallel to the second line of force, and a line is then drawn parallel to the first one from the end of the second one. The diagonal of the parallelogram that is formed starting at the common **application point** of the two forces is the **resultant force**.

PARALLELOGRAM RULE

$\vec{F_1}$

$\vec{R}$

$\vec{F_2}$

If two forces applied to a point have the same **direction** and **line of motion**, the resulting force follows the same line of motion and direction, and its **intensity** is the sum of the two intensities.

If two forces follow the same line of motion but have opposing directions, the resulting force will also have the same line of motion, the direction of the greater force, and an **intensity** that is the difference between the two intensities.

Introduction

Forces and Their Effects

Motion

Energy

Heat

Fluids

Wave Motion

Sound

Optics

Electricity

Matter

Inside Matter

Mixtures

Pure Substances

Chemical Changes

Alphabetical Subject Index

FORCES CONCURRENT AT ONE POINT

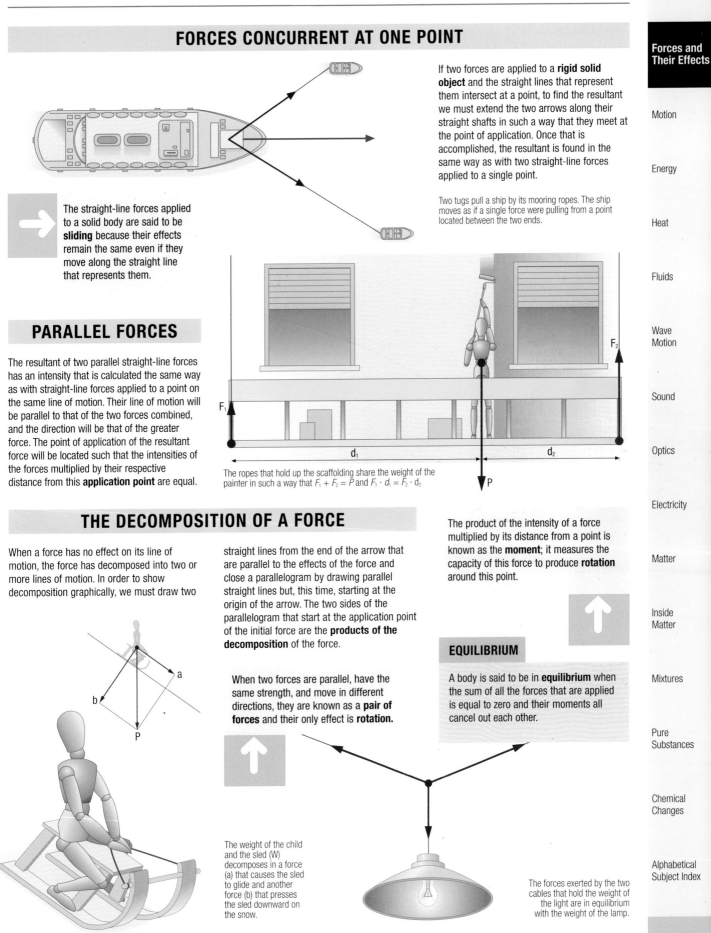

If two forces are applied to a **rigid solid object** and the straight lines that represent them intersect at a point, to find the resultant we must extend the two arrows along their straight shafts in such a way that they meet at the point of application. Once that is accomplished, the resultant is found in the same way as with two straight-line forces applied to a single point.

Two tugs pull a ship by its mooring ropes. The ship moves as if a single force were pulling from a point located between the two ends.

The straight-line forces applied to a solid body are said to be **sliding** because their effects remain the same even if they move along the straight line that represents them.

PARALLEL FORCES

The resultant of two parallel straight-line forces has an intensity that is calculated the same way as with straight-line forces applied to a point on the same line of motion. Their line of motion will be parallel to that of the two forces combined, and the direction will be that of the greater force. The point of application of the resultant force will be located such that the intensities of the forces multiplied by their respective distance from this **application point** are equal.

The ropes that hold up the scaffolding share the weight of the painter in such a way that $F_1 + F_2 = P$ and $F_1 \cdot d_1 = F_2 \cdot d_2$

THE DECOMPOSITION OF A FORCE

When a force has no effect on its line of motion, the force has decomposed into two or more lines of motion. In order to show decomposition graphically, we must draw two straight lines from the end of the arrow that are parallel to the effects of the force and close a parallelogram by drawing parallel straight lines but, this time, starting at the origin of the arrow. The two sides of the parallelogram that start at the application point of the initial force are the **products of the decomposition** of the force.

When two forces are parallel, have the same strength, and move in different directions, they are known as a **pair of forces** and their only effect is **rotation.**

The weight of the child and the sled (W) decomposes in a force (a) that causes the sled to glide and another force (b) that presses the sled downward on the snow.

The product of the intensity of a force multiplied by its distance from a point is known as the **moment**; it measures the capacity of this force to produce **rotation** around this point.

EQUILIBRIUM

A body is said to be in **equilibrium** when the sum of all the forces that are applied is equal to zero and their moments all cancel out each other.

The forces exerted by the two cables that hold the weight of the light are in equilibrium with the weight of the lamp.

FORCES AT A DISTANCE

The action of a force does not always come from **contact** between two bodies, one that produces the force and the other that is subjected to it. Other types of forces are produced between an **agent** and a **recipient** over extremely varied distances ranging from the vast distances that exist between two heavenly bodies to the different components of an **atom's nucleus**, including the forces created by an **electrical** or **magnetic field** at relatively short distances.

Equilibrium exists in the universe thanks to the forces of attraction among heavenly bodies.

THE LAW OF UNIVERSAL GRAVITATION

This law, formulated by Isaac Newton, relates the forces of **attraction** that exist between two bodies independently of the medium that separates them. The force by which two **masses** are attracted is directly proportional to those masses and inversely proportional to the square of the **distance** between them.

→ The constant of proportionality, known as the **universal constant of gravitation,** is so small (6.67×10^{-11} N·m²/kg², that is, 0.000,000,000,066.700 N·m²/kg²) that these forces are perceptible only when one or both masses are tremendously large, such as the earth.

GRAVITY

The **intensity of the gravitational field** at any point, normally referred to as gravity, is the force with which a heavenly body attracts the unit of mass located at that point. So gravity on the surface of a heavenly body depends on the **mass** and the body's **radius**.

UNITS OF GRAVITY

SYSTEM	UNIT
SI	N/kg
CGS	dyn/g
ST	kp/t.u.m.*

1 N/kg = 100 dyn/g
*technical unit of mass

WEIGHT

Weight is the **force** with which a heavenly body attracts a **mass** located inside its gravitational field. Thus, the weight of a body depends on its mass, the mass of the heavenly body, and the distance from the center of the heavenly body to the object.

On earth, gravity is much stronger than on the moon. This means that the weight of a suitcase or our body, for example, is much higher on earth than on the moon.

The weight of a mass m is equal to $w = m \cdot g$. Since its mass m is invariable, its weight depends on g. On the surface of the earth, g has an average value of 9.816 N/kg.

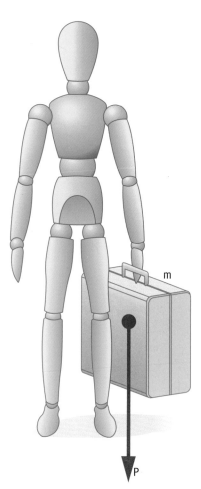

m

P

MAGNETIC FORCES

When certain **minerals,** such as **magnetite,** come close to a metal like **iron, nickel,** or **cobalt,** forces of attraction or repulsion come into play. These are called magnetic forces because the magnetite has created a **magnetic field.**

A magnet attracts
metal thumbtacks.

ELECTROSTATIC FORCES

Electrical charges exert forces of attraction or repulsion on one another. The area in space where the effects of an electrical charge are noticeable is called the **electrical field.**

When a comb is rubbed, it attracts
tiny bits of paper via electrostatic
forces.

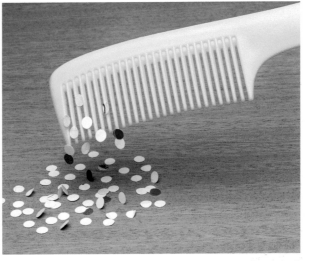

ELECTROMAGNETIC FORCES

An electrical current can produce a **field** referred to as **electromagnetic.** It is similar to a magnetic field. It exerts forces on electrical charges, magnetic fields, and bodies that contain any of those elements referred to as **ferromagnetic** (iron, cobalt, and nickel).

An electromagnet
attracts the bell ringer
intermittently.

NUCLEAR FORCES

These are tremendously intense forces of attraction that are responsible for the union of the particles in the nucleus of an atom and also for the great energy that is released when a **nucleus** is split apart. These forces are perceptible only at the infinitesimal distances that separate those particles.

The "wind" that is felt when you pass in front of a computer screen, even when it is turned off, is due to an electrical field produced by the tube that imparts an electrical charge to the screen's protective panel.

Motion

Energy

Heat

Fluids

Wave
Motion

Sound

Optics

Electricity

Matter

Inside
Matter

Mixtures

Pure
Substances

Chemical
Changes

Alphabetical
Subject Index

SIMPLE MACHINES

People have long understood that a lot of work is difficult to accomplish using only the strength of their arms. They used their ingenuity and advances in technology to conceive of a series of simple utensils to modify forces. These modifications may involve intensity, line of motion, and direction. An increase in intensity will always entail a reduction in trajectory as an inevitable consequence.

THE LEVER

The lever is a simple machine that modifies the **intensity** of forces and, therefore, the motion involved. Three basic elements are involved:

- **Effort force:** A point from which an outside force is exerted.
- **Support point** or **fulcrum**: the pivot point of the machine.
- **Load force**: the force to be overcome.

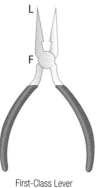

First-Class Lever

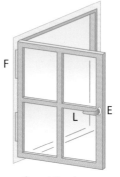

Second-Class Lever

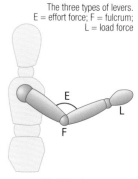

The three types of levers.
E = effort force; F = fulcrum;
L = load force

Third-Class Lever

Second-class levers always increase the strength of the force. Third-class levers always reduce the force. First-class levers increase or reduce the force depending on the location of the fulcrum.

TYPES OF LEVERS

Levers can be of three different types depending on the location of their three elements.

- **First class**: The fulcrum occupies a central position between the effort force and the load force. Example: pliers.
- **Second class**: The load force is located between the effort force and the pivot point. Example: a door.
- **Third class**: The effort force is in a central position between the load force and the fulcrum. Example: the human forearm.

The oar used by this gondolier in Venice is a second-class lever.

Even though the oars of a boat appear to be first-class levers, they really belong to the second type. The pivot point is the blade in the water, the effort force is located at the end of the oar held by the boater, and the load force is communicated to the boat through the oarlock.

TYPES OF LEVERS

Roman Scale

Load Force

Effort Force

Axis or Fulcrum

Nutcracker

Load Force

Effort Force

Axis or Fulcrum

Pliers

Load Force

Effort Force

Axis or Fulcrum

PULLEYS

The pulley is a simple machine that consists of a rope and one, two, or more grooved wheels for the rope. Pulleys reverse the direction of the force and can increase the strength of the force if there is more than one wheel. When a pulley has just one wheel, it is called a **single pulley**. If it has two wheels, it is described as a fixed and runner pulley. If it has a greater number of wheels, it is commonly called a block and tackle.

TYPES OF PULLEYS

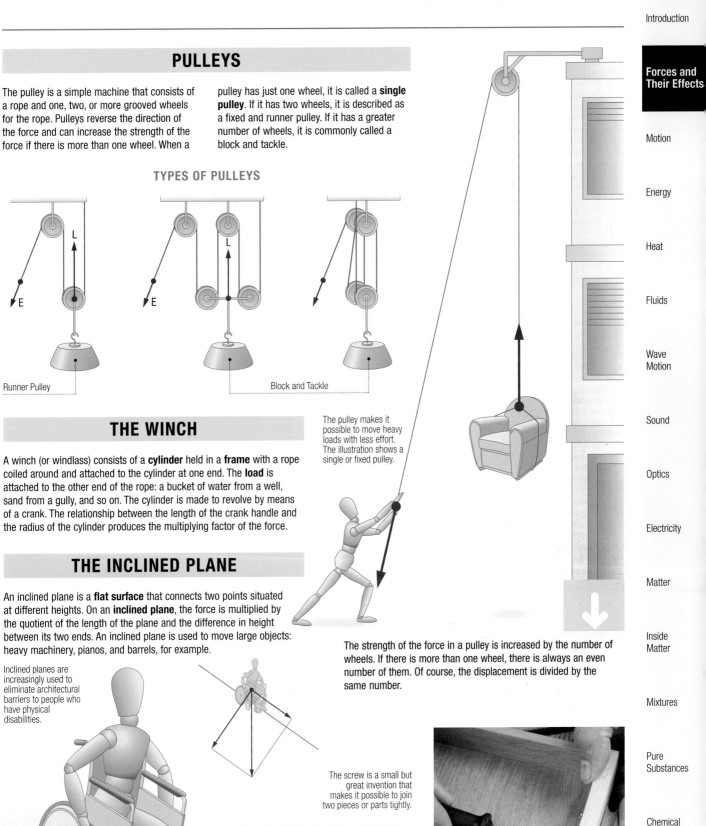

Runner Pulley

Block and Tackle

The pulley makes it possible to move heavy loads with less effort. The illustration shows a single or fixed pulley.

THE WINCH

A winch (or windlass) consists of a **cylinder** held in a **frame** with a rope coiled around and attached to the cylinder at one end. The **load** is attached to the other end of the rope: a bucket of water from a well, sand from a gully, and so on. The cylinder is made to revolve by means of a crank. The relationship between the length of the crank handle and the radius of the cylinder produces the multiplying factor of the force.

THE INCLINED PLANE

An inclined plane is a **flat surface** that connects two points situated at different heights. On an **inclined plane**, the force is multiplied by the quotient of the length of the plane and the difference in height between its two ends. An inclined plane is used to move large objects: heavy machinery, pianos, and barrels, for example.

Inclined planes are increasingly used to eliminate architectural barriers to people who have physical disabilities.

The strength of the force in a pulley is increased by the number of wheels. If there is more than one wheel, there is always an even number of them. Of course, the displacement is divided by the same number.

The screw is a small but great invention that makes it possible to join two pieces or parts tightly.

THE SCREW

The screw is a simple machine that consists of a **cylinder** covered by a **spiral groove** that changes the direction of the force and multiplies its strength according to its diameter and the distance between the grooves.

Introduction

Forces and Their Effects

Motion

Energy

Heat

Fluids

Wave Motion

Sound

Optics

Electricity

Matter

Inside Matter

Mixtures

Pure Substances

Chemical Changes

Alphabetical Subject Index

REFERENCE SYSTEMS

When we study a motion, we need to measure **distances** and **times**. In doing these measurements, we need to have the proper **measuring devices** and fixed **references** to form the basis of those **measurements**. This idea seems very simple. In reality, though, finding references that will guarantee accurate measurements in this ever-changing world in which we live is very difficult if not impossible.

WHEN DOES MOTION OCCUR?

We can be sure that motion occurs when, at any instant, we can indicate the **location** and the **velocity** of the **moving object**, predict these same characteristics at a specific **time**, or predict the time that the moving object will take to reach a certain position or travel at a certain velocity.

The palm tree located at the equator travels at the astonishing velocity of 1,017 mph (1,667 km/h).

A body that we consider to be at rest on the ground travels around 24,000 miles (40,000 km) every day due merely to the earth's rotation.

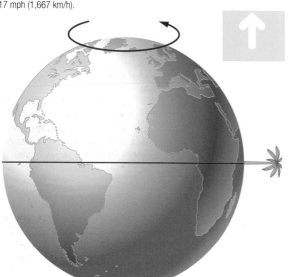

The passengers on this train who have looked at their timetable are informed of the train's motion since they know where it will be at any given moment.

THE RELATIVITY OF MOTION

All motion is **relative**. In the entire universe, there is no such thing as a place that occupies a fixed space. As a result, when studying motion, we consider a part of the universe to be at **rest**, and we can do measurements based on that supposition. If you are on a train traveling at high speed and are seated in a car, a passenger who is walking in the aisle will appear to you to be moving at a moderate speed. However, to an **observer** on the platform, the same person will be moving at a speed very similar to that of the train.

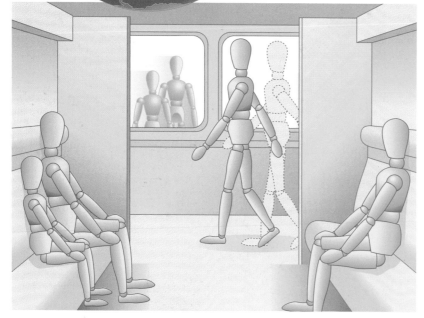

The man outside the train sees the passenger walking in the aisle move quickly past his eyes.

Introduction

Forces and
Their Effects

Motion

Energy

Heat

Fluids

Wave
Motion

Sound

Optics

Electricity

Matter

Inside
Matter

Mixtures

Pure
Substances

Chemical
Changes

Alphabetical
Subject Index

REFERENCE POINT

When an object is moving along a **straight line,** distances can be measured from a reference point. This point is considered point zero. **Positive** distances are measured from it in one direction and **negative** distances in the opposing direction.

REFERENCE SYSTEM

When motion occurs on a **plane** or in **space**, we need reference axes for measuring two or three distances, depending on the case, to locate the object at any moment.

In order to locate a point in space, three measurements are needed. They exist between that point and each of the three **Cartesian axes**.

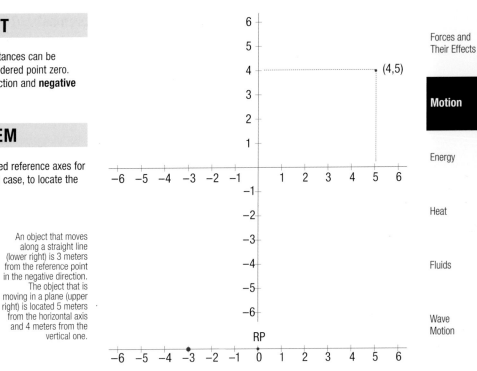

An object that moves along a straight line (lower right) is 3 meters from the reference point in the negative direction. The object that is moving in a plane (upper right) is located 5 meters from the horizontal axis and 4 meters from the vertical one.

RP

RELATIVE VELOCITY

If we study the motion of a body based on a system that is moving in the same direction, the **absolute velocity** of the object would be the sum of its velocity and that of the system. If the system is moving in the opposite direction, the absolute velocity of the object is the difference between the two velocities.

THE PRINCIPLE OF RELATIVITY

The distance traveled and time are two **absolute quantities**. This means that they do not depend on the velocity of the moving object or on the velocity of the observer. This statement is true at moderate velocities. However, at very high velocities, such as the **speed of light**, time and space cease to be absolute quantities and depend on that speed; a **relativistic contraction** of time and space occurs. At the speed of light, both time and space equal zero.

According to the theory of special relativity, the speed of light in a vacuum is 186,280 miles per second (300,000 km/s), independent of the light source. This quantity is one of the **universal constants.**

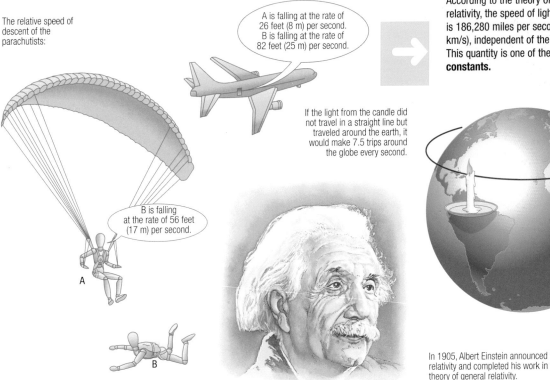

The relative speed of descent of the parachutists:

A is falling at the rate of 26 feet (8 m) per second. B is falling at the rate of 82 feet (25 m) per second.

B is falling at the rate of 56 feet (17 m) per second.

A

B

If the light from the candle did not travel in a straight line but traveled around the earth, it would make 7.5 trips around the globe every second.

In 1905, Albert Einstein announced the theory of special relativity and completed his work in 1915 with the theory of general relativity.

ELEMENTS THAT CHARACTERIZE MOTION

Imagine that you have to describe a movie to a friend—a movie whose name you cannot remember. What will you do? Imagine that you will provide a detailed description of the most important elements of the movie: the plot, the actors and actresses, the technical equipment, and interesting details so that only one movie would have all the mentioned characteristics. When describing motion, we have a series of characteristics we can use to describe a specific motion.

KINEMATICS

Kinematics is the part of **mechanics** that studies the motion of objects. Kinematics does not deal with the **causes** of the motion or its eventual **effects**.

Kinematics studies the motion of the skier without consideration of the fact that he is being pulled along by a kite.

POSITION

Position is the location that the moving object occupies at any given moment. The position is specified in distances measured according to the **reference system**. For example, if we say, "John was near the foot of his bed," anyone who is familiar with John's house will know perfectly well where John was at that moment.

Trajectory

Motion

Distance Traveled

When a snail moves, it marks its trajectory with its shiny slime

TRAJECTORY

When an object moves, it passes through successive positions. The line drawn through all these **positions** is known as the **trajectory.**

Star

10 Light-years
(5.88 trillion miles/94,608,000,000,000 km)

Earth

DISPLACEMENT

Displacement is the **distance** between two positions. This is not to be confused with **distance traveled,** which is the distance between two positions measured on the **trajectory.** We have to move in order to go from our home to school. The displacement is the straight-line distance between the two places. However, the distance traveled depends on the streets that each person chooses.

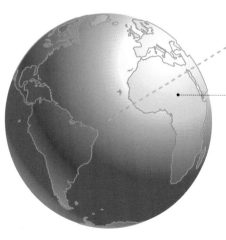

In astronomy, the distances between heavenly bodies are measured in light-years.

THE LIGHT-YEAR

The light-year is used in measuring huge displacements or distances. A light-year is the distance that a ray of light travels in one year: 186,280 miles (300,000 km) per second · 3,600 seconds per hour · 24 hours a day · 365 days per year = 5.88 trillion miles (9,460,800,000,000 km).

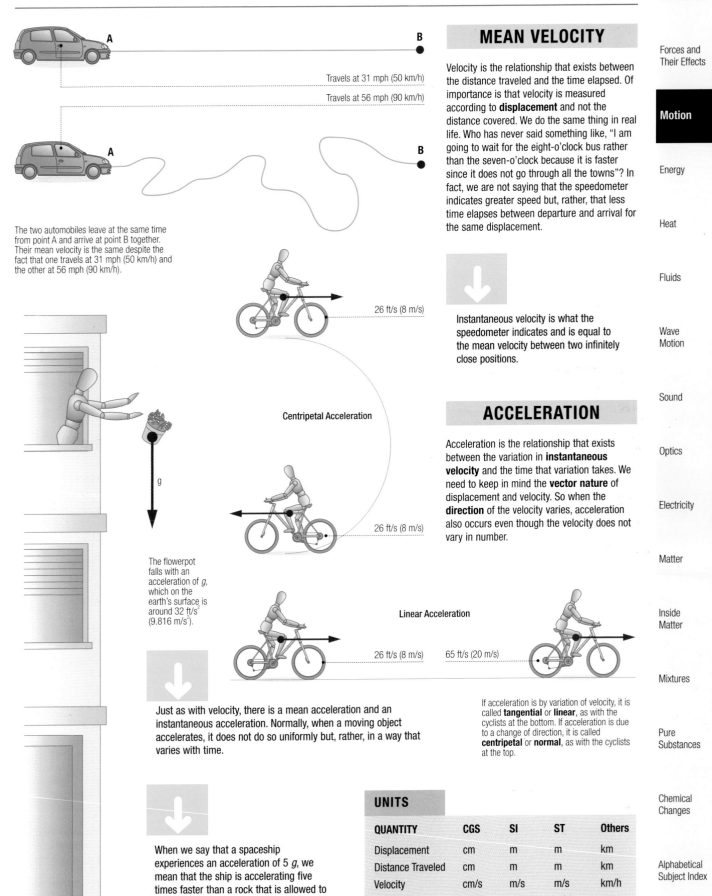

Forces and
Their Effects

Motion

Energy

Heat

Fluids

Wave
Motion

Sound

Optics

Electricity

Matter

Inside
Matter

Mixtures

Pure
Substances

Chemical
Changes

Alphabetical
Subject Index

Travels at 31 mph (50 km/h)

Travels at 56 mph (90 km/h)

The two automobiles leave at the same time from point A and arrive at point B together. Their mean velocity is the same despite the fact that one travels at 31 mph (50 km/h) and the other at 56 mph (90 km/h).

MEAN VELOCITY

Velocity is the relationship that exists between the distance traveled and the time elapsed. Of importance is that velocity is measured according to **displacement** and not the distance covered. We do the same thing in real life. Who has never said something like, "I am going to wait for the eight-o'clock bus rather than the seven-o'clock because it is faster since it does not go through all the towns"? In fact, we are not saying that the speedometer indicates greater speed but, rather, that less time elapses between departure and arrival for the same displacement.

Instantaneous velocity is what the speedometer indicates and is equal to the mean velocity between two infinitely close positions.

26 ft/s (8 m/s)

Centripetal Acceleration

26 ft/s (8 m/s)

g

The flowerpot falls with an acceleration of *g*, which on the earth's surface is around 32 ft/s² (9.816 m/s²).

ACCELERATION

Acceleration is the relationship that exists between the variation in **instantaneous velocity** and the time that variation takes. We need to keep in mind the **vector nature** of displacement and velocity. So when the **direction** of the velocity varies, acceleration also occurs even though the velocity does not vary in number.

Linear Acceleration

26 ft/s (8 m/s)   65 ft/s (20 m/s)

Just as with velocity, there is a mean acceleration and an instantaneous acceleration. Normally, when a moving object accelerates, it does not do so uniformly but, rather, in a way that varies with time.

If acceleration is by variation of velocity, it is called **tangential** or **linear**, as with the cyclists at the bottom. If acceleration is due to a change of direction, it is called **centripetal** or **normal**, as with the cyclists at the top.

When we say that a spaceship experiences an acceleration of 5 *g*, we mean that the ship is accelerating five times faster than a rock that is allowed to drop freely.

UNITS

QUANTITY	CGS	SI	ST	Others
Displacement	cm	m	m	km
Distance Traveled	cm	m	m	km
Velocity	cm/s	m/s	m/s	km/h
Acceleration	cm/s²	m/s²	m/s²	

TYPES OF MOTION

As we might suppose, there are countless types of motion. The study of motion involves everything from the slow, wobbly advance of an ant to the dramatic and rapid displacement of a jet engine. In order to facilitate the task, some typical motions have been identified to give a good idea of the calculations done. Real motions normally do not follow the predetermined ones precisely, but they can be very similar.

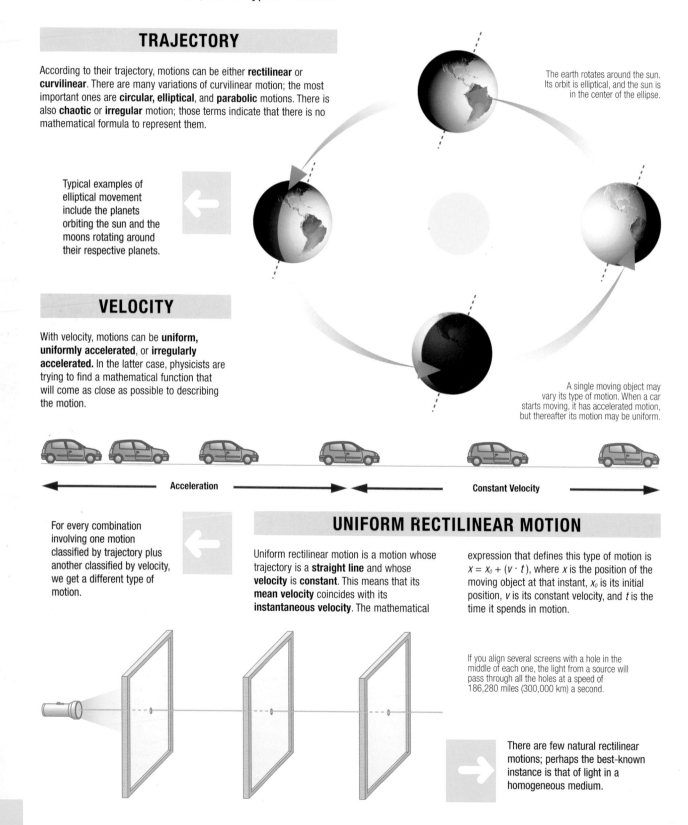

TRAJECTORY

According to their trajectory, motions can be either **rectilinear** or **curvilinear**. There are many variations of curvilinear motion; the most important ones are **circular, elliptical**, and **parabolic** motions. There is also **chaotic** or **irregular** motion; those terms indicate that there is no mathematical formula to represent them.

The earth rotates around the sun. Its orbit is elliptical, and the sun is in the center of the ellipse.

Typical examples of elliptical movement include the planets orbiting the sun and the moons rotating around their respective planets.

VELOCITY

With velocity, motions can be **uniform, uniformly accelerated**, or **irregularly accelerated.** In the latter case, physicists are trying to find a mathematical function that will come as close as possible to describing the motion.

A single moving object may vary its type of motion. When a car starts moving, it has accelerated motion, but thereafter its motion may be uniform.

Acceleration **Constant Velocity**

For every combination involving one motion classified by trajectory plus another classified by velocity, we get a different type of motion.

UNIFORM RECTILINEAR MOTION

Uniform rectilinear motion is a motion whose trajectory is a **straight line** and whose **velocity** is **constant**. This means that its **mean velocity** coincides with its **instantaneous velocity**. The mathematical expression that defines this type of motion is $x = x_0 + (v \cdot t)$, where x is the position of the moving object at that instant, x_0 is its initial position, v is its constant velocity, and t is the time it spends in motion.

If you align several screens with a hole in the middle of each one, the light from a source will pass through all the holes at a speed of 186,280 miles (300,000 km) a second.

There are few natural rectilinear motions; perhaps the best-known instance is that of light in a homogeneous medium.

UNIFORMLY ACCELERATED RECTILINEAR MOTION

Forces and
Their Effects

Motion

Energy

Heat

Fluids

Wave
Motion

Sound

Optics

Electricity

Matter

Inside
Matter

Mixtures

Pure
Substances

Chemical
Changes

Alphabetical
Subject Index

In this type of motion, the trajectory is a straight line, but the velocity varies uniformly. This means that the **acceleration** is **constant**. In order to define this motion, we need two mathematical expressions that are closely related to one another:
$x = x_0 + (v_0 \cdot t) + (\frac{1}{2} \cdot a \cdot t^2)$ and $v = v_0 + (a \cdot t)$.

A graphic representation of time versus velocity of a uniformly accelerated motion with negative acceleration.

SYMBOLS AND UNITS

Position	Initial Position	Velocity	Initial Velocity	Time
x	x_0	v	v_0	t
m	m	m/s	m/s	s

THE MOTION OF FALLING BODIES

Objects that are subject to the action of a **gravitational field** fall toward the center of the planet with a **uniformly accelerated rectilinear** motion. The acceleration of this motion is the intensity of the gravitational field expressed in m/s². Thus, when we fall down, we do so at an acceleration of 32.205 ft/s² (9.816 m/s²). In reality, things do not happen exactly that way because of **friction** with the air.

Since friction with the air increases with velocity, there is a **terminal velocity**. This is reached when the force of friction is equal to **mass**. From that moment on, the motion is uniform since the velocity does not vary.

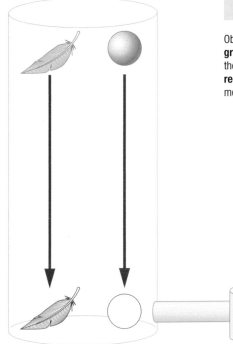

THE BEHAVIOR OF A MOVING OBJECT

Vacuum Pump

In a vacuum, a ball of lead and a feather fall at the same velocity and hit the bottom at the same instant.

A hang glider lands harmlessly because of the shape of the parachute, whose terminal velocity is quite low.

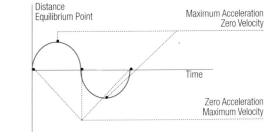

Distance
Equilibrium Point

Maximum Acceleration
Zero Velocity

Time

Zero Acceleration
Maximum Velocity

Vibratory motion is said to be a **periodic motion** because for every unit of time, known as a **period**, the object passes through the same point, at the same velocity, and in the same direction.

VIBRATORY MOTION

Have you ever seen a doll suspended from the roof of a baby's room by a soft spring? If we pull on it, it keeps moving back and forth for a long time. This motion is a **vibratory** motion known as a **simple harmonic**. This motion is rectilinear since its trajectory is a **segment** of a straight line. Its **velocity** will be at its **maximum** when it passes through the center of its trajectory and **zero** at the ends. So it is accelerating, but not uniformly accelerating, since its **acceleration** is at a **maximum** at the ends (it even changes direction of motion) and **zero** at the midpoint of its travel.

UNIFORM CIRCULAR MOTION

The trajectory of this motion has a **circumference** and its velocity is constant. There are two different types of velocity in this type of motion, and they are closely related to one another: **linear velocity** and **angular velocity**. The two velocities are related by the **radius** of the circumference. The mathematical expressions that define this motion are the following: for linear quantities, $v = l/t$, and for angular quantities, $\omega = \varphi/t$, where l is the arc traveled; φ is the angle of rotation, v is the linear velocity, ω is the angular velocity, and t is the time.

ACCELERATION

As we might suppose, there are two types of **acceleration**: **linear** and **angular**. Angular acceleration is the relationship between the variation of angular velocity and elapsed time. Linear acceleration has two components. **Tangential acceleration** is the relationship between the variation of velocity in intensity and time. **Normal or centripetal acceleration** is calculated mathematically by $a_c = v^2/R$, where a_c is centripetal acceleration, v is velocity, and R is radius. This is the acceleration we experience when traveling in an automobile that turns without varying its speed.

A record turntable revolves at an angular velocity of 45 rpm (revolutions per minute)

A radian equals the angle at the center of a circle subtended by an arc whose length is the same as the radius of the circle.

RADIAN (IN RED)

The speedometer of an automobile indicates linear velocity. We can calculate the tangential acceleration by dividing the average velocity by the time.

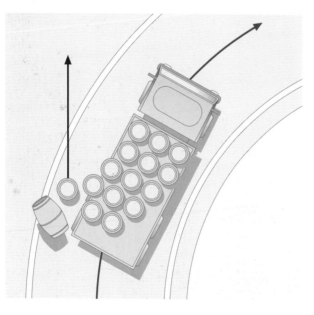

WHAT PRODUCES ACCELERATION?

When an object experiencing acceleration supports another (or several) object(s), some inertial forces become apparent. These forces tend to make the two objects take on different motions. In fact, the object that is not subject to acceleration tries to follow its own path with a uniform rectilinear motion. So when the stone that rotates within a slingshot is released, it fires along the direction of the cord at that instant.

When the truck turns, the barrels in the bed move toward the outside of the curve since they continue in a straight line.

Introduction

Forces and
Their Effects

Motion

Energy

Heat

Fluids

Wave
Motion

Sound

Optics

Electricity

Matter

Inside
Matter

Mixtures

Pure
Substances

Chemical
Changes

Alphabetical
Subject Index

RELATIONSHIPS BETWEEN LINEAR AND ANGULAR QUANTITIES

Have you ever played the following game? Five or six children hold hands and stretch out in a straight line. When the youngster at the end goes around in a circle, the others try to go around at their own angular velocity. Accomplishing this is almost impossible for the youngster on the end. Why? Because the greater the distance from the center of the turn, every **angle of turn** requires a greater distance. To do a complete revolution, the youngster in the center barely has to take a few steps. However, the last youngster at the end has to travel $2 \cdot \pi \cdot R$, where R is the distance that separates the two youngsters. The entire circumference has 2π radians, so the relationship between the angle and the **arc** is the radius. Thus:

- $l = \varphi \cdot R$
- $v = \omega \cdot R$
- $a = \alpha \cdot R$

as long as angular quantities are measured in radians.

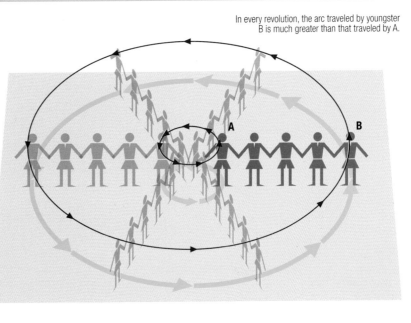

In every revolution, the arc traveled by youngster B is much greater than that traveled by A.

Vibratory and pendular motion differ only in their paths. Therefore, pendular motion also has a period.

UNITS AND SYMBOLS

QUANTITY	SYMBOL	UNIT
Length of Arc	l	m
Arc	φ	rad
Linear Velocity	v	m/s
Angular Velocity	ω	rad/s
Linear Acceleration	a	m/s^2
Angular Acceleration	α	rad/s^2

PENDULAR MOTION

We can create a **pendulum** by hanging a large object from a wire and attaching the end of that wire to a support. When the object is moved from its **equilibrium position,** it takes on a motion referred to as pendular. It is a circular motion because its trajectory is a **circumferential arc**. The motion is not uniform because its velocity continually varies. It is also not uniformly accelerated because its acceleration also varies. Now look closer. At the ends of the motion, the pendulum stops and **changes its direction** of motion. Its velocity is greatest when it passes through its equilibrium point, and at that instant, its acceleration is zero.

The period of pendular motion has been used for many years in marking time on clocks.

CHARACTERISTICS OF PERIODIC MOTIONS

Periodic motions have two very closely related characteristics: the **period** and the **frequency**. The period is the time that the moving object requires for one complete **oscillation**, that is, passing two consecutive times through the same point in the same direction. The frequency is the number of oscillations that the moving object completes in a unit of time.

THE COMPOSITION OF MOTIONS

If you play basketball, it never occurs to you to throw the ball directly at the basket in a straight line. Of course, it would not go through the hoop. We know only too well that when we throw an object, it does not follow a straight trajectory but, rather, travels along a parabola.

Why? If there were no force, the object would follow a straight trajectory at a constant velocity. Its weight, though, causes it to fall at an acceleration equal to gravity. The parabolic motion comes from the sum of the two motions.

HORIZONTAL MOTION

Imagine a diver running on the 5-meter board. When the diver reaches the end, he takes off horizontally at a certain velocity. If the **forces of friction** did not exist, this velocity would be constant. Suppose for a moment that friction does not exist. From the instant that his feet no longer touch the board, he would start to fall in a **uniformly accelerated** motion. On a coordinate graph, his position (x, y) on the axes at any given moment would be $x = v \cdot t$ and $y = h - (\frac{1}{2} \cdot 9.816 \cdot t^2)$. Recall that he has no initial vertical velocity and that when he leaves the board his position is zero horizontally and h (the height of the board). By replacing the time with specific values, we can determine the **position** of the diver at any instant. When he hits the water, his position is zero.

OBLIQUE MOTION

A study of oblique motion is very similar to that of a horizontal motion. First, we break down the **initial velocity** into the sum of the **horizontal velocity** and the upward **vertical velocity**. Then we consider the motion to be the sum of **two motions perpendicular** to one another: a vertical motion (a uniformly accelerated motion) and a horizontal motion (a uniform motion). The equations are:
$x = v_x \cdot t$; $y = (v_y \cdot t) - (\frac{1}{2} \cdot 9.816 \cdot t^2)$. The height at which the throw is made is presumed to be zero.

From a single point, it is possible to reach the same target using two different angles.

THE COMPOSITION OF TWO PERPENDICULAR VIBRATORY MOTIONS

It is interesting to see that the sum of two perpendicular vibratory motions can produce very different trajectories. The variables that establish the trajectory are the **frequency**, the **amplitude**, and the **initial position** of the two motions.

DISTANCE AND MAXIMUM HEIGHT

The distance is how far the projectile travels, measured horizontally.

The maximum value of y is referred to as the maximum height.

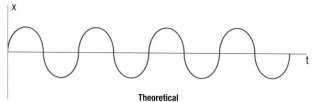

Theoretical

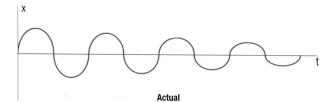

Actual

Real vibratory motions diminish in amplitude with time. This fact is used in making shock absorbers for automobiles.

Introduction

Forces and
Their Effects

Motion

Energy

Heat

Fluids

Wave
Motion

Sound

Optics

Electricity

Matter

Inside
Matter

Mixtures

Pure
Substances

Chemical
Changes

Alphabetical
Subject Index

SAME FREQUENCY AND MAXIMUM EXPANSION

If two vibratory motions are perpendicular to one another, possess the same frequency, and both begin motion at the greatest amplitude, the result will be an **oblique vibratory motion** whose angle depends on the amplitude of the initial motions.

If the two motions have the same amplitude, the angle will be 45°.

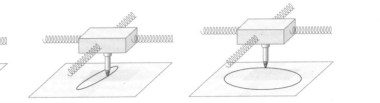

The marker indicates the path produced by the vibrating springs.

If the **amplitude** of two motions is the same and the initial position is one of equilibrium while that of the other is farthest from equilibrium, the resulting movement is circular.

SAME FREQUENCY AND DIFFERENT EXPANSION

When an object is subjected to two simultaneous vibratory motions that are perpendicular to one another and have the same frequency, we may discover in studying the two motions individually that the initial position of the object is different in each motion. In one of them, it is the point farthest from the equilibrium point, and in the other, it is a point close to the equilibrium point. The motion is an **elliptical motion**. The farther the initial position of the second motion is from its equilibrium point, the greater the **eccentricity** of the ellipse described by the moving object.

DIFFERENT FREQUENCY

When an object is subjected to two simultaneous motions that are perpendicular to one another and have different frequencies, the result is a periodic motion with a strange trajectory known as the **curves of Lissajous**.

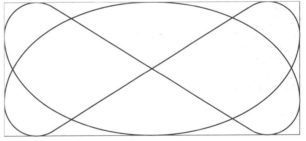

The curves of Lissajous can be produced by combining two motions that are perpendicular to one another.

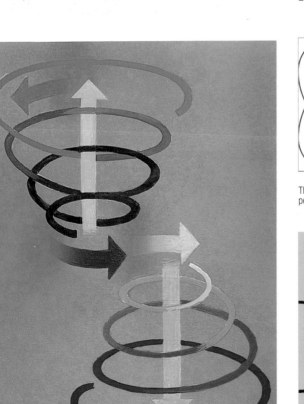

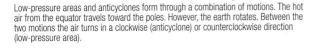

Low-pressure areas and anticyclones form through a combination of motions. The hot air from the equator travels toward the poles. However, the earth rotates. Between the two motions the air turns in a clockwise (anticyclone) or counterclockwise direction (low-pressure area).

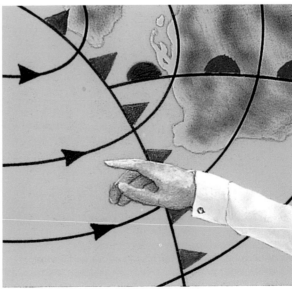

FORCE AND MOTION

The definition of force indicates that there is a relationship between force and motion. Our common sense tells us, sometimes incorrectly, that all motion is due to a force. We can picture this misconception by imagining a steel disk sliding over a horizontal, icy surface. We cannot see the end of this motion. Does our imagination deceive us? No. The forces are related to the variations in the motion but not to the motion itself.

NEWTON'S PRINCIPLES

Dynamics, the study of the relationships between forces and motion, is based on three of Newton's principles, or the principles of dynamics:

- The principle of **inertia**: an object cannot alter its state of **rest** or **motion** by itself.

- The **fundamental** principle: **mass** is the relationship that exists between the **force** received by an object and the **acceleration** that it acquires: $F = m \cdot a$.

- The principle of **action** and **reaction**: For every action, there is an equal and opposite reaction. The reaction has the same **intensity** but travels in the opposite direction and has a different point of application.

A spaceship in outer space travels with uniform, straight-line motion since in the absence of force, there is no acceleration.

In order to slow down an object in motion, a force is needed in the direction opposite that of the velocity.

The English physicist, mathematician, and astronomer Isaac Newton (1642–1727).

The sum of all the forces applied to the shopping cart equals its mass times the acceleration that it experiences: $F - F_r = m \cdot a$. If there is no acceleration, the force F equals friction.

When a person jumps onto the dock, the boat moves in accordance with the principle of action and reaction.

The application point of the action force is on the object that experiences the action of this force. The application point of the reaction force is on the object that carries out the force.

THE ACCELERATION OF GRAVITY

When we **let go of an object**, it falls. The force that makes the object fall is its **weight**, which we have seen is $W = m \cdot g$. It is easy to deduce that if $F = m \cdot a$ and the weight is the force that makes objects fall, $a = g$. So the **intensity of the gravitational field** and the **acceleration** of gravity are equal in magnitude but not in units.

THE NEWTON

A Newton is the force that is applied to a kilogram of mass so that it acquires an acceleration of 1 meter per second squared: $1\ N = 1\ kg \cdot m/s^2$.

Introduction

Forces and
Their Effects

Motion

Energy

Heat

Fluids

Wave
Motion

Sound

Optics

Electricity

Matter

Inside
Matter

Mixtures

Pure
Substances

Chemical
Changes

Alphabetical
Subject Index

OBJECTS IN FREE FALL

Dynamics has explained the acceleration of gravity. So the falling motion of objects is a **uniformly accelerated rectilinear** motion whose acceleration is the acceleration of **gravity**. This fact allows us to calculate the velocity at which an object will arrive if allowed to fall freely through the air (ignoring friction): $v = g \cdot t$, or to calculate the height of a tower according to the time that an object takes to fall: $h = g \cdot t^2/2$.

The chest gains speed
as it falls.

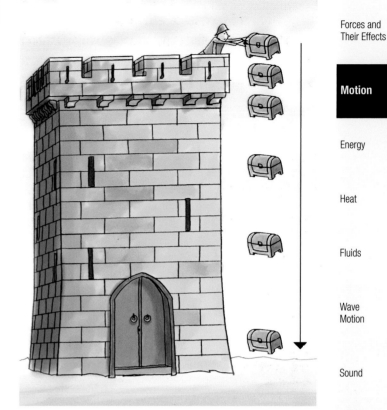

THE FORCE OF FRICTION

Friction is a **reaction** force. When we have an object on a surface and we **slide it**, from a microscopic point of view the irregularities of the **object** and the **surface** engage one another. When we push on the object, it pushes against the surface, which pushes the object in the **opposite direction**, thereby complicating our task.

Although it is hard to believe, motion exists because of friction. Think about it: we are able to walk because of friction between the ground and our shoes.

ROPE TENSION

Tension is the **force** that a rope experiences in carrying out a specific action. Look at two cases: when tension works to move something horizontally on a surface and when the motion is vertical.

- When we drag an object **horizontally,** we have to overcome two forces at the same time: **friction** with the underlying surface and the **force of inertia:**
$T = F_f + (m \cdot a)$.

- Look at an elevator cable. When the elevator begins to go up, if we consider **friction** to be zero, the cable has to support the **weight** of the elevator and its load plus impart **acceleration**. Thus, the tension is
$T = (m \cdot g) + (m \cdot a)$.

FORCES AND A CRANE

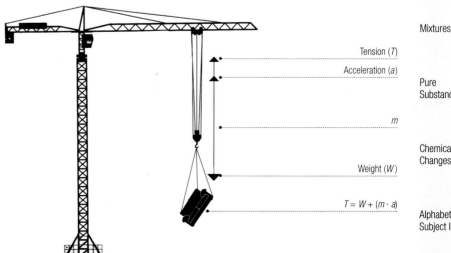

Tension (T)
Acceleration (a)

m

Weight (W)

$T = W + (m \cdot a)$

When an elevator that is going up slows down, the cable has to support a tension equal to the weight of the elevator plus the force of inertia. However, since that is negative, the tension is less than the weight.

WORK, POWER, AND ENERGY

Everything that we do in life involves work. In order to do work, we take precautions: we eat as well as possible, we drink, we sleep, and so forth. In other words, in order to carry out work, we store up energy. Whenever it becomes necessary, this energy is transformed into work, and this in turn is transformed into a more degraded type of energy, usually heat. We say that energy is degraded when it becomes harder to put to use.

PHYSICAL WORK

In physics, the word **work** means the result of the intensity of a **force** multiplied by the **displacement** produced in the line of motion of that force. We must make a distinction between this word and the way it is used colloquially in the sense of **exertion**. So we have to keep in mind that three things have to come together for work to be produced:

- There has to be a force.

- The force has to displace the object.

- The line of motion of the force and the displacement must not be perpendicular to one another.

Humans eat in order to accumulate energy and carry out physical work.

Work can be defined mathematically in the following way: $W = F_x \cdot d_x$. X stands for horizontal force.

The man using a wheelbarrow exerts a force that has two effects: F_1, which displaces it (accomplishing work), and F_2, which raises it (accomplishing no work).

1 kw · h is work equivalent to 3,600,000 joules, which is about the amount of work involved in raising 792,000 pounds (360,000 kg) 1 yard (1 m).

UNITS OF POWER AND WORK

CGS	SI	ST	Others
erg	J (joules)	kpm (kilopond meter)	kw · h (kilowatt hour)
erg/s	W (watts)	kpm/s	HP (horsepower)

POWER

What kind of work can a **machine** perform? The answer is easy: any kind. However, what we do not know is the **time** needed to do it. The relationship between the work performed and the time consumed in performing it is called **power**: $P = W/t$. Thus, the motor of a truck is more powerful than that of an automobile because the truck can produce much more work in the same time.

The motor of a large truck develops much more power than that of a common automobile.

Introduction

Forces and
Their Effects

Motion

Energy

Heat

Fluids

Wave
Motion

Sound

Optics

Electricity

Matter

Inside
Matter

Mixtures

Pure
Substances

Chemical
Changes

Alphabetical
Subject Index

ENERGY

Energy is the term applied to everything that can be converted to **work**, whether because of its situation (**potential**) or state of motion (**kinetic**). Work is produced during the change in form of the energy. The water in a **mill pond**, a compressed **spring**, a **rock** at the top of a mountain, and a disconnected **voltaic battery**, for example, all have potential energy. When the water runs out of the mill pond, the tension on the spring is released, the rock falls, and the battery is connected; they all possess kinetic energy.

Potential energy in batteries makes it possible for certain portable objects to function independently.

TYPES OF ENERGY

There are countless types of energy. We merely need to think of everything that is capable of producing **motion**. The most important energy types include **mechanical, chemical, electrical, thermal, radiant,** and **nuclear**. These types of energy are **transformable** from one form to another. Consider, for example, that in a nuclear power plant, the nuclear energy is transformed into heat energy, which is turned into mechanical energy, which in turn produces electrical energy. Once the energy is in the grid, it is transformed into any form that the consumer desires.

In most cases, energy that is transformed into work takes the form of thermal energy that disperses in the air.

TYPES OF ENERGY

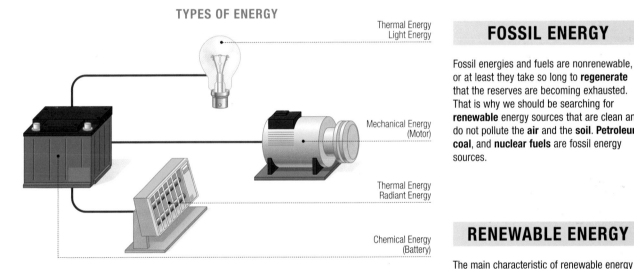

Thermal Energy
Light Energy

Mechanical Energy
(Motor)

Thermal Energy
Radiant Energy

Chemical Energy
(Battery)

FOSSIL ENERGY

Fossil energies and fuels are nonrenewable, or at least they take so long to **regenerate** that the reserves are becoming exhausted. That is why we should be searching for **renewable** energy sources that are clean and do not pollute the **air** and the **soil**. **Petroleum, coal,** and **nuclear fuels** are fossil energy sources.

RENEWABLE ENERGY

The main characteristic of renewable energy sources is that they are produced continually and are practically **inexhaustible**; plus, they pose no problems with **pollution**. For now, these energy sources are not a substitute for fossil fuels. However, their use needs to increase on a daily basis so we will have a clean planet with the needed energy. Renewable energies include energy from the sun (**solar**), the wind (**eolian**), the earth's internal heat (**geothermal**), the potential of water stored at certain heights (**hydraulic**), and others.

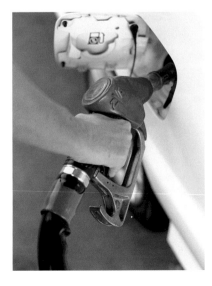

At left, the gasoline that we use in our cars comes from petroleum, a fossil energy source that will be exhausted within a few more decades.

Photovoltaic cells capture clean and inexhaustible energy from the sun.

MECHANICAL ENERGY

A hammer's capacity to accomplish work, the destructive power of a thrown rock, and the cocked wooden throwing arm of a catapult all are examples of objects endowed with a type of energy termed mechanical. In order to realize their capacity for work, they must be put into motion. Consider that in giving examples of mechanical energy, we have named some objects. That fact allows us to infer that mechanical energy is what a mass possesses because of its state.

Ancient catapults used deformational potential energy.

POTENTIAL ENERGY

Potential energy is the energy that objects have because of their **situation**, and it is not transformed directly into work. Potential mechanical energy may be **deformational** or **gravitational**. In order to understand the former, you merely need to think of an old-style clock that needs to be wound every day. As the small wheel is turned, a spring is compressed. This spring changes shape as it recovers its former shape and is responsible for the **motion** of the hands.

GRAVITATIONAL POTENTIAL ENERGY

When we raise a pick in order to drive it into the ground, the higher we raise it the deeper it will dig into the earth. In raising the pick, we effect work. The return on this work comes when the pick hits the ground. The **potential energy** that the pick has at its highest point is known as **gravitational potential energy** because it uses the action of **gravity**. Potential gravitational energy can be expressed mathematically as $E_p = m \cdot g \cdot h$.

The absolute value of potential energy is usually not known. The calculated value is the **difference** in energy that exists between two points located at different **heights**.

Many times we use the action of gravity in swinging a hammer.

KINETIC ENERGY

Kinetic energy is the type of **mechanical energy** that objects have because of their **motion**. Kinetic energy is calculated mathematically as $E_k = \frac{1}{2} \cdot m \cdot v^2$. Since every object has *mass*, if an object is in motion, it has *kinetic energy*. If a **force** acts on an object **at rest** or **in motion** and is not perpendicular to the trajectory, it will accomplish work. If the force contributes to velocity, this work is transformed into kinetic energy. If the force works in the opposite direction, the energy, that is, the velocity, is reduced.

When a father pushes his child on a tricycle, he is performing work that is transformed into kinetic energy.

Introduction

Forces and
Their Effects

Motion

Energy

Heat

Fluids

Wave
Motion

Sound

Optics

Electricity

Matter

Inside
Matter

Mixtures

Pure
Substances

Chemical
Changes

Alphabetical
Subject Index

QUANTITY OF MOTION

When a **force** acts on an object for a given **time,** it imparts a certain velocity to it. The product of the force multiplied by the time it acts is known as the **mechanical impulse,** and the product of the **mass** multiplied by the **velocity** it has is called the **quantity of motion**.

In kinetic energy, since velocity is squared, when a car doubles its speed, the danger (kinetic energy) is multiplied by four.

The bowling ball on the right has four times the kinetic energy of the ball on the left; however, it has twice the quantity of motion.

VARIATION IN THE QUANTITY OF MOTION

The mechanical impulse is equal to the variation in the quantity of motion it produces: $F \cdot t = (m \cdot v) - (m \cdot v_0)$.

ELASTIC IMPACTS

If we watch the **collision** between two billiard balls, we see that the contact is instantaneous. Since they are made of a very hard substance, there is hardly any deformation. This is an example of an elastic impact. No **loss** of **kinetic energy** or **quantity of motion** occurs in this type of impact.

The white cue ball has quantity of motion and kinetic energy. These quantities do not vary after contact, but they are shared between the cue ball and the eight ball.

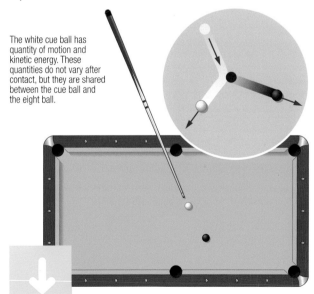

No type of impact in real life is perfectly elastic. A loss of velocity, that is, of kinetic energy, always occurs, no matter how small.

INELASTIC IMPACTS

Impacts that occur between two objects that remain **joined** after the shock are termed inelastic or plastic shocks. One example is shot pellet that strikes a piece of wood set up on a fence rail. In this case, the quantity of motion is also preserved, but the kinetic energy experiences a noticeable loss. The greater the **work** performed in the **deformation** that produced the union, the greater the loss in kinetic energy.

When the striker kicks the soccer ball, the contact with the ball is practically elastic. However, when the ball reaches the goalie, the contact is inelastic.

Many real-life shocks are elastic ones, but there are also quite a few **inelastic** ones.

THE CONSERVATION OF ENERGY

When we hoist a bucket of water up to the fourth floor, we expend energy and perform work. Is the energy consumed equal to the work performed? Theoretically, it may be. In reality, though, there is always a loss due to friction. Here is another question we might ask: Is the energy expended lost in the performance of the work? No. The expended energy has been converted into potential energy that is imparted to the bucketful of water and certainly into the heat that is produced in the pulley.

THE CONSERVATION OF MECHANICAL ENERGY

If no forces are exerted on a system, the **mechanical energy** is conserved. Mechanical energy means the sum of the **kinetic energy** plus the **potential energy**. Look at an example. A rock held at the edge of a cliff has mechanical energy equal to its potential energy since it is not in motion. Now drop the rock, and suppose there is no **friction** with the air. The height of the rock decreases, but its velocity increases in such a way that the sum $(m \cdot g \cdot h) + (\frac{1}{2} \cdot m \cdot v^2)$ is constant. When the rock hits the water, its altitude is zero, and all its energy has been converted into **kinetic** energy.

If we throw an object straight up, the kinetic energy that we give it is converted to potential energy at the **highest point**.

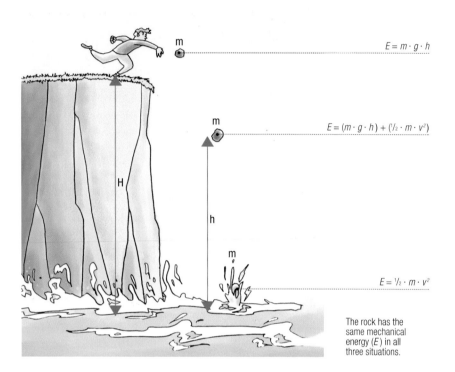

$$E = m \cdot g \cdot h$$

$$E = (m \cdot g \cdot h) + (\frac{1}{2} \cdot m \cdot v^2)$$

$$E = \frac{1}{2} \cdot m \cdot v^2$$

The rock has the same mechanical energy (E) in all three situations.

CALCULATING VELOCITIES AND HEIGHTS

The principle of conservation of energy helps us calculate the maximum height that an object launched vertically will attain and the velocity that an object launched vertically or allowed to fall freely in a vacuum will attain. In so doing, we need only to calculate the **initial** and the **final** mechanical energy, including the unknown quantities represented by their symbols. Ultimately, the two types of energy will be equal. For now, we need only to find the value of the desired quantity: E (initial) = E (final).

WORK AS AN ENERGY MODIFIER

If forces are exerted on an object that we are examining, the mechanical energy will not be conserved. The force or forces applied move along with the object, and consequently, they perform **work** on it. This work is added to the **mechanical energy** if it contributes to the motion (**positive work**) or is subtracted if it goes against the motion (**negative work**).

The mortar launches the fireworks to an altitude where they have potential energy equal to the kinetic energy they had in the muzzle of the cannon.

The work of moving the three books from the lower to the higher shelf (positive work) increases the potential energy.

There is one force from which we can never free ourselves: **friction**. Sometimes we can disregard it if it is very minor.

In trying to hold back a dog, a person performs negative work that reduces the speed of the animal along with its kinetic energy.

THE CONSERVATION OF ENERGY

The **conservation of energy** is a universal principle: energy is neither created nor destroyed, simply transformed. As described on the previous page, mechanical energy is not conserved when a **force** is applied to the object we are examining. The force carries out **work**; this work originated from some other source of **energy**. Besides, whenever work is carried out, some energy is always lost as heat or converted into another form of energy. Therefore, the principle of conservation of energy tells us that the energy used when examining the object plus the energy produced by the work performed on the object equals the resulting energy plus the heat released during performance of the work.

Energy

Heat

Fluids

Wave
Motion

Sound

Optics

Electricity

The temperature of the nail increases when the work performed by the hammer is converted to heat.

Usually the heat given off when work is performed passes into the atmosphere and from there to the lakes and seas that act as graveyards for energy. Only a little part of this energy is renewable. The photo shows the Liffey River as it passes through Dublin, Ireland.

IS ENERGY REALLY INVARIABLE?

In 1905, Albert Einstein announced the equivalency of mass and energy. This means that mass can be transformed into energy. This energy can be calculated mathematically on the basis of the mass destroyed: $E = m \cdot c^2$. E is the energy produced, m is the mass destroyed, and c is the speed of light (186,200 miles per second/300,000 km/s). In both physics and in classical chemistry, the energy absorbed or given off amounts to thousands of joules. Since the speed of light is such a vast number, the variation in the mass is so small that it cannot be measured. This variation can be measured, though, in nuclear reactions.

Inside
Matter

Mixtures

Pure
Substances

Chemical
Changes

Alphabetical
Subject Index

THE CONSERVATION OF MASS AND ENERGY

Throughout any physical or chemical process, the sum of the mass and energy remains invariable.

The mass of the nucleus of helium gas is slightly lower than that of the two protons plus the two neutrons because a lot of energy is given off when the nucleus forms.

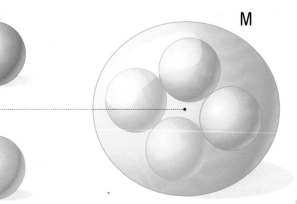

m

M

Neutron

Nucleus of Helium Gas

M

Proton

HEAT, A SOURCE OF ENERGY

Have you ever noticed how water appears cloudy as it heats up? At first you can see the impurities moving around in a random pattern inside the water. As the water heats up (taking on more heat), the visible solids increase the speed of their motion until the water begins to boil, at which time our observation becomes imperfect. This fact indicates to us that in order to store up heat, matter increases the kinetic energy of its particles. When it cools down, this energy is passed on to the object that absorbs the heat.

THE TRANSFORMATION OF WORK INTO HEAT

When an automobile travels on a highway at a certain speed, it has **kinetic energy**. If it **brakes** and stops, a force has been exerted on it. **Work** has been performed that cancels out all its kinetic energy. Where is this work? If we touch the brakes, we see that they have become very hot. All the work has dissipated in the form of **heat**.

THE TRANSFORMATION OF HEAT INTO WORK

The heat stored in an object (a **heat source**) dissipates by traveling to another object at a lower temperature (a **heat sink**). During this heat exchange, **work** may be performed in a **thermal machine**. We must recall that not all of the **transferred heat** is converted into work, and some heat will always pass into the atmosphere or to a refrigerant.

When a match is scratched on the striker, it produces the heat needed to make the match burst into flame.

The heat that leaves the heat source is shared between the heat sink and the atmosphere. The latter occurs after the work is produced.

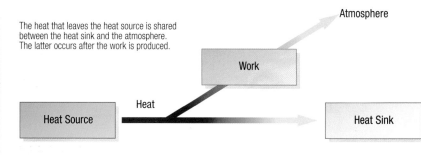

All the work produced can turn into thermal energy, but only a part of the thermal energy can be transformed into work.

THERMAL MACHINES

The simplest thermal machine is a tin can with a perfectly round hole plugged with a cork. We merely need to place this can into a double boiler and wait for the **temperature** of the air inside the can to rise. When the pressure inside the can reaches a certain level, the **cork pops out**, resulting in work because both force and displacement occur. If we did this experiment inside a container where the temperature was 212°F/100°C, the cork would never come out since the inside and outside temperatures would be equal.

A simple thermal machine.

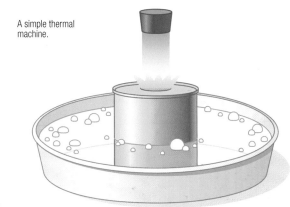

HOUSEHOLD THERMAL MACHINES

There are several thermal machines in the kitchen of any house. Their purpose is of course not the performance of work, but work happens just the same. A **coffee pot** is one example. With heat, the water rises from one chamber to a higher one. Another example is a **pressure cooker**. When the temperature increases, the steam pressure in the upper part of the pot increases and the pressure valve turns, allowing the steam to escape to the outside, where the temperature is lower.

Introduction

Forces and
Their Effects

Motion

Energy

Heat

Fluids

Wave
Motion

Sound

Optics

Electricity

Matter

Inside
Matter

Mixtures

Pure
Substances

Chemical
Changes

Alphabetical
Subject Index

THE OUTPUT OF A THERMAL MACHINE

In a machine, the heat source receives a quantity of heat Q at a temperature T_1 and gives up a quantity of heat Q_2 to the heat sink at a temperature of T_2, which is less than T_1. Only the difference between Q_1 and Q_2 can be converted into work. The quotient $R = (T_1 - T_2)/T_1$ is known as the **theoretical performance**. Only if the temperature of the heat sink is 32°F/0°C is the output one.

T represents the absolute temperature. It equals degrees in Celsius plus 273.16. The temperature of −273.16°C is known as absolute zero.

The **practical output** is the quotient of the **work** produced and the energy obtained by **completely burning** the fuel.

Steam engines are still used in some industries for driving large machines for extended periods of time.

STEAM ENGINES

At some time, we have all seen those old locomotives that belch smoke skyward and are enveloped in white steam. They are old, but they were the first thermal machines that existed for the purposes of transportation. A **boiler** was used to raise the water temperature to form steam **under pressure** and at **high temperature**. This steam seeks a way toward the outside by pushing on a piston, which leads to the rotation of the wheels. An ingenious method makes the steam push the piston alternately in both directions.

SCHEMATIC DRAWING OF A STEAM BOILER

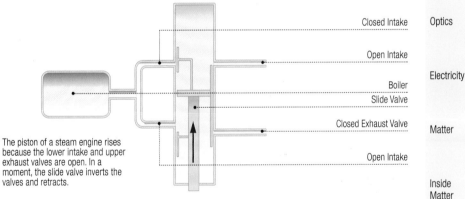

Closed Intake

Open Intake

Boiler

Slide Valve

Closed Exhaust Valve

Open Intake

The piston of a steam engine rises because the lower intake and upper exhaust valves are open. In a moment, the slide valve inverts the valves and retracts.

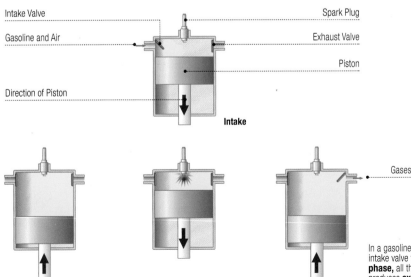

Intake Valve

Gasoline and Air

Direction of Piston

Spark Plug

Exhaust Valve

Piston

Intake

Compression   **Ignition and Expansion**   **Exhaust**

Gases

INTERNAL COMBUSTION ENGINES

These are **motors** that make direct use of the heat produced by burning a **fossil fuel** to produce **work.** They may be **two-cycle** motors used in some motorcycles; **four-cycle** motors, which have long been used in most automobiles; or **diesel injection** motors, which function at such a high temperature that they do not need combustion initiators. Presently, **turbo motors** are being used, which are supercharged by fuel and air injection pumps.

In a gasoline-powered four-cycle engine, in the **intake phase**, only the intake valve for the mixture of gas and air is open. In the **compression phase,** all the valves are closed. When the gasoline burns with the air, it produces **expansion**, and all the valves are still closed. In the **exhaust phase,** the exhaust valve opens to let out the combustion gases.

THE EFFECTS OF HEAT

Aside from the production of work through thermal machines, in our daily lives we know of other phenomena that are based on heat: the heating of objects, the temperature, changes in physical state, and others. However, experience shows us that not all materials react the same way to heat. For example, although water takes a long time to heat up, other materials such as silver spoons quickly reach high temperatures.

HEAT PROPAGATION

Heat can migrate from a heat source to a heat sink through three procedures:

- **Contact**: Heat is the vibration of particles. A particle in a state of vibration communicates its motion to neighboring particles. An iron bar with one end in a fire soon grows hot over its entire length.

- **Convection:** Hot liquids and gases are **less dense** than cold ones. Therefore, hot **gases** and **liquids rise** and are replaced by cold ones, which become heated, and so forth. For example, a pot of soup that is removed from the burner is hotter at the top than at the bottom.

- **Radiation**: This involves electromagnetic **waves** (infrared rays). The heat from the sun reaches us through this means since it uses no **solid structure** on its journey toward the earth.

The heat from the frying pan reaches the chef's hand by contact through the spoon.

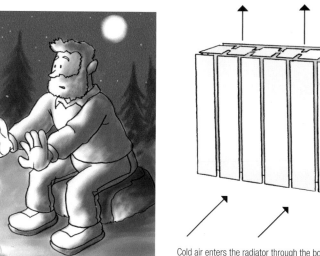

When we warm our hands by a campfire, we are taking advantage of heat transfer through radiation.

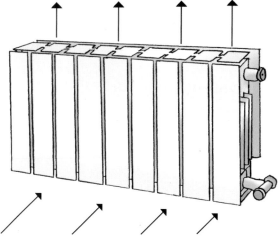

Cold air enters the radiator through the bottom and leaves through the top after being heated, thereby setting up circulation through convection.

EXPANSION

In the summer, as the temperature increases, we can see the mercury of a thermometer rise. Why? Materials take up more **volume** when the temperature is higher. This phenomenon is known as **volumetric expansion**. The volume occupied by an object is as follows: $V = V_0 (1 + \gamma t)$, where V_0 is the volume that it took up at 0°C; γ is the coefficient of volumetric expansion, which depends on the expanding solid or liquid (but not gas); and t is the temperature in degrees Celsius.

Approximately every 200 yards/meters is a gap in the railroad tracks to allow for expansion during the summer.

Introduction

Forces and
Their Effects

Motion

Energy

Heat

Fluids

Wave
Motion

Sound

Optics

Electricity

Matter

Inside
Matter

Mixtures

Pure
Substances

Chemical
Changes

Alphabetical
Subject Index

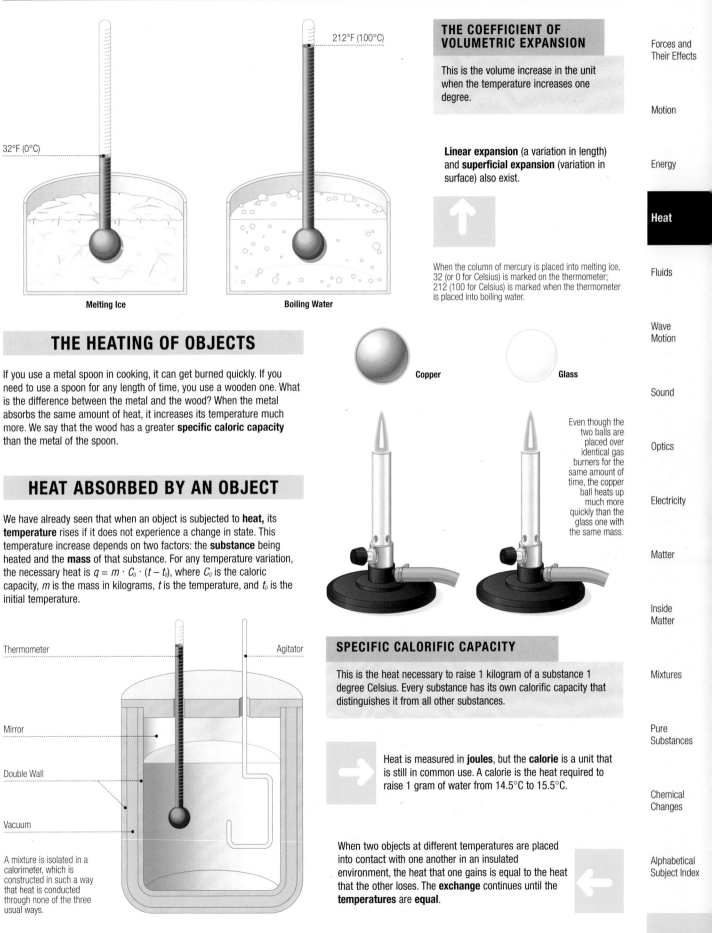

212°F (100°C)

32°F (0°C)

Melting Ice

Boiling Water

THE COEFFICIENT OF VOLUMETRIC EXPANSION

This is the volume increase in the unit when the temperature increases one degree.

Linear expansion (a variation in length) and **superficial expansion** (variation in surface) also exist.

When the column of mercury is placed into melting ice, 32 (or 0 for Celsius) is marked on the thermometer; 212 (100 for Celsius) is marked when the thermometer is placed into boiling water.

Copper

Glass

Even though the two balls are placed over identical gas burners for the same amount of time, the copper ball heats up much more quickly than the glass one with the same mass.

THE HEATING OF OBJECTS

If you use a metal spoon in cooking, it can get burned quickly. If you need to use a spoon for any length of time, you use a wooden one. What is the difference between the metal and the wood? When the metal absorbs the same amount of heat, it increases its temperature much more. We say that the wood has a greater **specific caloric capacity** than the metal of the spoon.

HEAT ABSORBED BY AN OBJECT

We have already seen that when an object is subjected to **heat,** its **temperature** rises if it does not experience a change in state. This temperature increase depends on two factors: the **substance** being heated and the **mass** of that substance. For any temperature variation, the necessary heat is $q = m \cdot C_0 \cdot (t - t_0)$, where C_0 is the caloric capacity, m is the mass in kilograms, t is the temperature, and t_0 is the initial temperature.

Thermometer

Agitator

Mirror

Double Wall

Vacuum

A mixture is isolated in a calorimeter, which is constructed in such a way that heat is conducted through none of the three usual ways.

SPECIFIC CALORIFIC CAPACITY

This is the heat necessary to raise 1 kilogram of a substance 1 degree Celsius. Every substance has its own calorific capacity that distinguishes it from all other substances.

Heat is measured in **joules**, but the **calorie** is a unit that is still in common use. A calorie is the heat required to raise 1 gram of water from 14.5°C to 15.5°C.

When two objects at different temperatures are placed into contact with one another in an insulated environment, the heat that one gains is equal to the heat that the other loses. The **exchange** continues until the **temperatures** are **equal**.

CHARACTERISTICS OF LIQUIDS

No one has any difficulty classifying materials according to their physical state. However, it is difficult to describe the properties that are common to all liquids, whose consistency distinguishes one liquid from another. Why is it important to know about liquids and their properties? The answer is in the abundance and the vital importance of liquids. One liquid, water, covers a major part of the earth and is a major component of living beings.

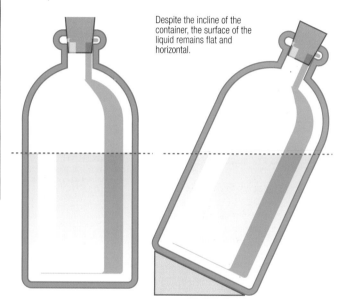

Liquids are **fluid**. This means that their particles are in constant motion, as with gases. In the case of liquids, though, the particles can only slide over one another.

WHAT IS A LIQUID?

We say that a material is liquid when it has no **shape** of its own and adopts the shape of the container that holds it, it is incapable of being **compressed**, and its top surface—as long as it is free—is **flat** and **horizontal**. When liquids are found in large areas, such as the water in the ocean or in lakes, the shape of the surface may be altered by atmospheric forces.

Despite the incline of the container, the surface of the liquid remains flat and horizontal.

FORCES OF COHESION

Most properties of liquids come from what is known as **forces of cohesion**. These forces are the ones that exist among particles of the same **liquid** and those that exist among the particles of the liquid and the **container walls.**

The particles of the liquid are attracted to one another and to the surface particles of the container.

LIQUIDS AND CONTAINERS

Depending on the container, liquids can be of two types: those that **moisten** the container and those that **do not moisten** it. Surely we have noticed that water droplets **flatten out** on a glass surface. In contrast, if the drops are of mercury, they retain their rather spherical shape. However, if the water droplets fall onto a new frying pan used for frying without oil, they take on a spherical shape; water moistens glass but not the pan.

Droplets assume a spherical shape in Teflon-coated pans.

Introduction

Forces and
Their Effects

Motion

Energy

Heat

Fluids

Wave
Motion

Sound

Optics

Electricity

Matter

Inside
Matter

Mixtures

Pure
Substances

Chemical
Changes

Alphabetical
Subject Index

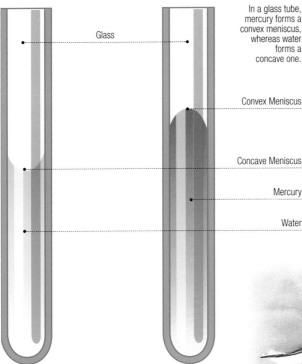

In a glass tube, mercury forms a convex meniscus, whereas water forms a concave one.

Glass

Convex Meniscus

Concave Meniscus

Mercury

Water

 Near the walls of a container, the free surface of liquids is **concave** if the liquid moistens the recipient and **convex** if the liquid does not moisten it. This concave or convex shape is known as a **meniscus**.

SURFACE TENSION

Have you ever placed a waxed pin onto the surface of a glass of water? If you exercise enough care, you see that even though the pin is very **dense**, it **floats** on the surface of the water. The surface of the water acts like a **taut membrane** that withstands the weight of the pin. If you add soap to the water, surface tension will form bubbles that splash water when they rise and evaporate.

The water glider is a hemipteran insect that walks on water using surface tension and its small body and long legs, which are not wet by the water.

CAPILLARITY

Why does wax **go up** the candlewick to meet the flame? The liquids that moisten the container walls, in this case the cotton of the wick, rise through the **tiny channels** created by the strands. Liquids that do not moisten are **rejected** by the same channels. In times of drought, wild plants provide water for themselves by means of capillarity. Deep water rises through dry ground to the roots. The same liquid can **go up** or **down** through a capillary, depending on the material from which the capillary is made.

Water rises in the thin glass tube (capillary action), but mercury does not reach the level of the outside.

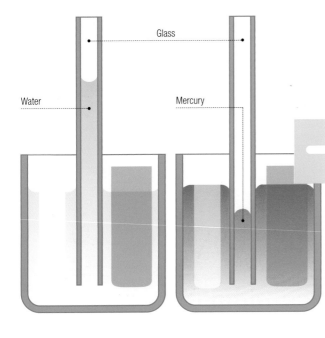

Glass

Water

Mercury

VISCOSITY

Suppose that we have a smooth, glass surface that forms an incline plane. At the top of this plane, we deposit a drop of water and, at the same moment, a drop of oil right beside it. The two drops have a contest. The drop that moves faster at the ambient temperature is the water droplet. So we say that the oil is more viscous. That means that the viscosity is the **force** that keeps the particles of the liquid from **sliding** against one another. Viscosity decreases significantly with a rise in **temperature.**

Many scientists consider **noncrystalline solids** such as wax and glass to be extremely viscous **liquids**.

Because of its viscosity, a drop of oil slides more slowly than a drop of water.

DROPS RELEASED AT THE SAME TIME

Drop of Oil

Drop of Water

THE THEOREMS OF PASCAL AND ARCHIMEDES

Liquids are transported, but they are also useful for transporting other things. Huge pipelines carry petroleum from one area to another and even go across countries, lakes, and deserts. Water has been used for transporting materials not only in ships but also by floating logs downstream in big timber operations and moving the rocks in primitive gold sluicing setups. Since it is such a valuable commodity, water is stored in large artificial lakes.

DENSITY

In what ways are water and mercury similar? Finding similarities is certainly more difficult than finding differences. One noteworthy difference is that given the same **volume** of each substance, their weight is very different. Since there is a **mass** that corresponds to every weight, we can say that they have very different mass. The **relationship** between mass and volume is known as **density:** $d = m/V$, which is measured in kg/m³ in the SI.

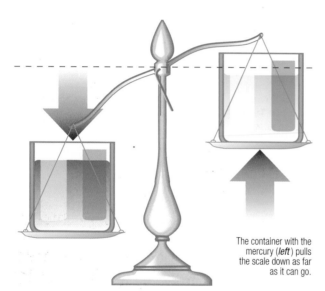

The container with the mercury (*left*) pulls the scale down as far as it can go.

If we divide weight by volume rather than mass by volume, we get specific weight. Although the density of an object is constant, the *specific weight* depends on the gravity of a particular area.

PRESSURE

When we walk on the snow, we note that we sink in to a greater or lesser degree depending on our footwear. If we wear snowshoes or skis, we sink in less. What has changed? Not the mass or the weight but the surface on which the weight is supported. The **relationship** between the **force** and the **surface** that supports it is called **pressure:** $p = F/S$. The unit in the SI is the **pascal** (Pa).

COMMUNICATING VESSELS

Liquids move from **greater** to **lesser** pressure. If we join two containers at the bottom, the **level** of the liquid in each will be the same even though the quantities may be very different as long as both containers have the same liquid. Of course, if one level were higher than the other, one point of the conduit that joins the two containers would be subject to different pressures, and the liquid would flow toward the side with the lower pressure until the two levels were equal.

Since skis have a large surface area, the skier does not sink into the snow.

The pressure that a liquid exerts on a point depends on the density of the liquid and the distance between that point and the free surface.

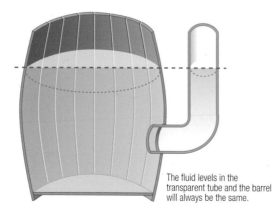

The fluid levels in the transparent tube and the barrel will always be the same.

Introduction

Forces and
Their Effects

Motion

Energy

Heat

Fluids

FLOTATION

Why does a cork float on water, and why does something made of iron not float? What property distinguishes the two and makes them behave so differently? The density of iron is 17,160 pounds (7,800 kg) per cubic meter; the density of cork is 880 pounds (400 kg) per cubic meter; the density of water is 2,200 pounds (1,000 kg) per cubic meter. When we **submerge** an object, it displaces a volume of liquid equal to its volume. The weight of iron is greater than that of the water it displaces since the water is less **dense**. However, the weight of the cork is less than that of the water it displaces since the cork's density is lower. Thus both the iron and the cork are pushed upward by the water, but only the cork **floats** because that pushing force exceeds the cork's weight.

PASCAL'S THEOREM

If **pressure** is applied to a liquid in equilibrium, the pressure is transmitted equally to **all points**.

THE PRINCIPLE OF ARCHIMEDES

Every **object submerged** in a liquid experiences an **uplifting force** equal to the **weight** of the liquid that it displaces.

APPLICATIONS OF PASCAL'S THEOREM

There are many applications for the phenomenon described by Blaise Pascal. One of them is the **hydraulic press**. You probably have seen how mechanics in repair shops raise automobiles several feet off the floor to work under them comfortably. The automobile is supported by a **large cylinder** while the mechanic or a small motor activates a **very small cylinder**. The pressure in both cylinders is the same. Even though the pressure is low in the small cylinder, a lot of pressure is produced in the large one, lifting the car.

Wave
Motion

Sound

Optics

Electricity

Matter

Inside
Matter

Mixtures

Pure
Substances

Chemical
Changes

Alphabetical
Subject Index

A powerful hydraulic piston (the yellow tube and the steel shaft) is used in raising the sections of this crane; it is based on Pascal's principle.

Brake Pedal

Cylinder

The force the driver applies by foot to the piston is transmitted to the two pistons of the brake cylinder, which act on the brake shoe. The brake shoe presses against the brake drum that is part of the car's wheel.

Double Cylinder

Brake Drum

Brake Shoe

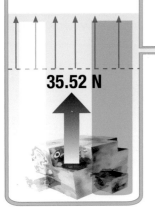

35.52 N

The volume of the object submerged in the glass and that of the liquid that has overflowed are the same. The object is subjected to an upward force equal to the weight of the liquid.

When an object floats, it always has some part that is submerged. The weight of the liquid that this part displaces equals the total weight of the object.

GASES

How could we not be interested in studying gases? Like most animals, we live submerged in gas just as fish live submerged in water. This gas provides us with the materials we need for the slow combustion that we call **respiration.** Air, like all gases, is difficult to measure. We cannot define the quantity of gas by volume alone; we need more information to know how much we are dealing with.

WHAT IS A GAS?

We refer to substances that display two negative requirements as gases:

- They have no shape of their own and take on the shape of their container.
- They have no volume of their own and likewise take on the volume of their container.

The first characteristic is the same as with liquids, with the difference that there is no free surface with gases. The second characteristic gives them a property referred to as **compressibility.**

THE PRESSURE-VOLUME RELATIONSHIP

We have all done the following experiment at some time: when using a bicycle pump, you block the valve with one finger and compress the piston. What happens? The pump still has the same amount of air (since we keep our finger in place), and yet the volume it occupies is much smaller. We have compressed the air.

Gases are fluid, just as liquids are, but gas particles move in all directions and are not in contact with one another.

If we block the valve on the pump and push on the cylinder, we compress the air that it contains.

The scent of a perfume quickly reaches every corner of a room.

ABSOLUTE TEMPERATURES

Gases, like all materials, expand as temperatures rise but with one special characteristic: all gases have the same coefficient of expansion, 1/273.15°C. When a gas is frozen, its volume decreases because its particles slow down. So at a temperature of −459.67°F (−273.15°C) all its particles have their minimum possible energy and its volume is practically equal to zero. This temperature is called *absolute* or **Kelvin.** As we have already mentioned concerning the effects of heat, the temperature in Kelvin is the temperature in degrees Celsius plus 273.15: $T_K = T_C + 273.15$.

BOYLE-MARIOTTE'S LAW

For the same mass of gas at a constant temperature, the product of multiplying pressure by volume is invariable: $P \cdot V = K$.

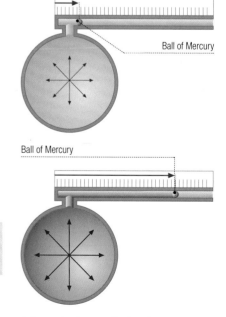

Ball of Mercury

Ball of Mercury

As temperature increases, the drop of mercury moves to make room for the gas in the reservoir.

Introduction

Forces and
Their Effects

Motion

Energy

Heat

Fluids

Wave
Motion

Sound

Optics

Electricity

Matter

Inside
Matter

Mixtures

Pure
Substances

Chemical
Changes

Alphabetical
Subject Index

GAY-LUSSAC'S LAWS

First: For the same quantity of gas at constant pressure, the volume is directly proportional to the absolute temperature: $V/T = K$.

Second: For the same quantity of gas at constant volume, pressure is directly proportional to the absolute temperature: $P/T = K$.

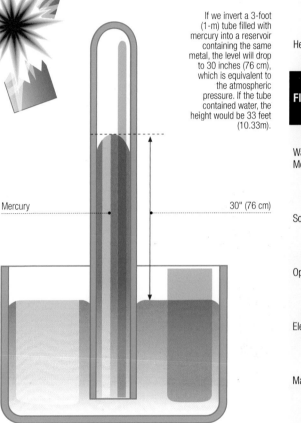

Pressurized aerosol cans that contain gas bear a warning label against exposure to sunlight since an increase in temperature could cause an explosion by an excessive increase in pressure.

ATMOSPHERIC PRESSURE

Atmospheric pressure is the **pressure** exerted by **air** at a point in the atmosphere. Atmospheric pressure varies with altitude because the layer of air is thicker at the surface of the earth than at the top of a mountain. At one time or another, we have all experienced the effects of this variation when we quickly went up or down a mountain in a car. We have realized that all of a sudden, our ears feel blocked, and they can even hurt.

THE EFFECTS OF ATMOSPHERIC PRESSURE

Atmospheric pressure will be perceptible whenever we enter an area that has a different pressure. Have you ever performed magic using a glass, water, and a piece of paper? Place a piece of nonporous paper on top of a glass completely full of water. Turn the glass over quickly with your hand; the water stays in place because it is supported by the paper. This is because the atmospheric pressure is greater than the pressure exerted on the paper by the water.

Atmospheric pressure keeps the paper and water in place when the glass is inverted.

Normal atmospheric pressure at the earth's surface is 1 atmosphere, which equals 101.325 kilopascals (kPa) or 1,013.25 millibars (mb).

When the air is removed from the oil can, atmospheric pressure crushes the can.

IDEAL GAS LAW

When we pump up the tires of a bicycle, fill a lighter with gas, or compress or expand a gas, we notice a change in temperature. It is very difficult to vary just two of the three variables: P, V, and T. So we need a mathematical equation that establishes the relationships among the three. This is the **ideal gas law** equation: $(P \cdot V)/T = K$.

If we invert a 3-foot (1-m) tube filled with mercury into a reservoir containing the same metal, the level will drop to 30 inches (76 cm), which is equivalent to the atmospheric pressure. If the tube contained water, the height would be 33 feet (10.33m).

Mercury

30" (76 cm)

Vacuum Pump

CHARACTERISTICS OF WAVES

There is no more fantastic natural spectacle than watching a storm at sea. From afar we see the huge waves course toward the beach and break onto the sand. The waves really move toward the shore, but does the water do the same? On a stormy day, it would not be strange to see a log floating on the water. If we carefully observe the motion of the log as it drifts, we see that it does not necessarily move in the direction of the waves. The log exhibits mostly a back-and-forth motion, a motion that we have already seen under the name of vibratory motion in preceding sections.

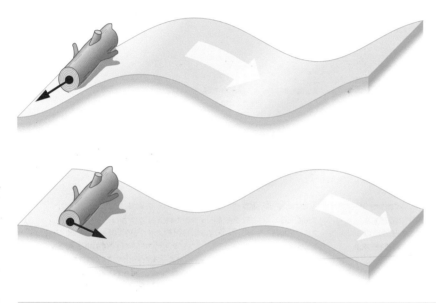

WHAT IS A WAVE?

A point in an elastic medium is subjected to **energy**. The point begins to vibrate. This **vibration** is communicated through contact to nearby points, where the vibration arrives after a slight **delay**. This delay depends on the distance from the point where the vibration began. Imagine that the vibration is energy and that the energy is transmitted, but the points—that is, the vibrating materials—are never **displaced.** Thus we see that the phenomenon can be broken down into two parts: a motion of materials that vibrate (**kinetic energy**) and the **propagation** of this motion to other particles.

Depending on the nature of the wave, the log will move in one direction or another.

TYPES OF WAVES

If we throw a rock into the calm water of a lake, we can see the water start to **oscillate vertically** where the rock hits as the initial wave communicates its motion **horizontally** to the water around it. This type of wave motion, in which the disturbance and the propagation are perpendicular to one another, is known as **transverse waves**. Now pretend to secure a coil spring at one of its ends. Next pull together some spirals at the other end of the spring and then release them. These spirals separate spontaneously, but in their **expansion**, the next set of spirals come together in succession. We see that in this motion, the disturbance and the propagation follow the **same line of motion**. These are known as **longitudinal waves.**

In transverse waves, crests alternate with depressions.

With longitudinal waves, there is a series of compressions and expansions.

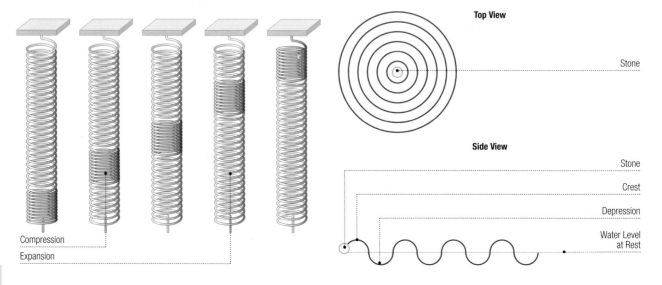

Compression
Expansion

Top View

Stone

Side View

Stone
Crest
Depression
Water Level at Rest

Introduction

Forces and
Their Effects

Motion

Energy

Heat

Fluids

**Wave
Motion**

Sound

Optics

Electricity

Matter

Inside
Matter

Mixtures

Pure
Substances

Chemical
Changes

Alphabetical
Subject Index

WAVELENGTH

Wavelength is the **distance** that exists between two consecutive points located the same distance from the **equilibrium point** (where they were before the start of the agitation) that are traveling at the same **velocity** and in the same **direction**. In ocean waves, wavelength is the distance between **two consecutive crests**. Wavelength is measured in meters and is represented by the Greek letter λ.

FREQUENCY AND PERIOD

Frequency is the **number of oscillations** in the motion in a given unit of time. The unit is the **hertz**, and the symbol is the Greek letter v.

Period is the **time** required for a particle to complete **one entire oscillation**. The unit is the second and the symbol is the letter T.

PROPERTIES OF WAVES

If a wave collides with an obstacle and bounces off, it is **reflected**. If the wave passes from one medium to another, as it changes velocity it **deflects**, a phenomenon known as **refraction**. If a wave reaches an obstacle that has an opening the size of the wavelength, the **opening** turns into a sender of another wavelength of the same frequency as the original one. This phenomenon is known as **diffraction**.

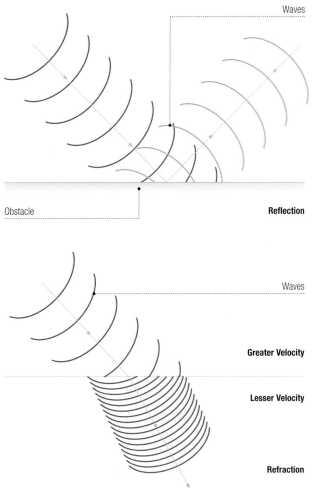

Waves

Obstacle   **Reflection**

Waves

Greater Velocity

Lesser Velocity

Refraction

WAVELENGTH AND AMPLITUDE

Wavelength

Amplitude

The maximum distance that a particle achieves relative to its equilibrium point is known as **amplitude**.

RELATIONSHIPS AMONG QUANTITIES

The more oscillations per minute, the less time required for each oscillation. This means that frequency and period are inversely proportional: $v = 1/T$. We also have to consider that in a homogeneous medium, the propagation of a wave follows a **uniform rectilinear motion**. We are already familiar with the equation for motion. Since when a wave moves we call the time required a **period**, we can write the equation as $v = \lambda/T$ or $v = \lambda \cdot v$, where v is the **propagation velocity**.

The propagation velocity of a wave depends only on the **medium** in which it propagates and not on its characteristics.

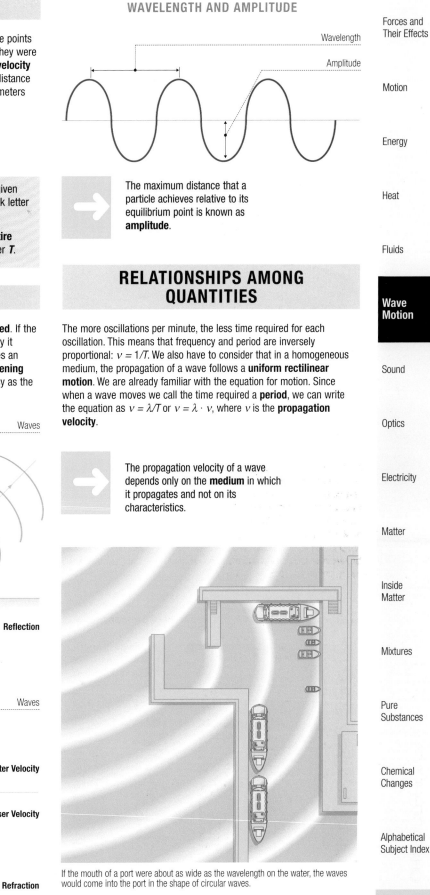

If the mouth of a port were about as wide as the wavelength on the water, the waves would come into the port in the shape of circular waves.

SOUND

Have you ever paused to consider how sounds reach our ears? Why do sounds often seem louder but distorted in an enclosed area? We will see that sound is nothing more than an undulatory phenomenon that originates, as in all cases, with a vibratory motion produced by impact or friction. Therefore, sound is one form that energy adopts to displace itself through a material medium: air, wood, iron, water, and so forth.

THE NATURE OF SOUND

There is no doubt about the **undulatory nature** of sound, but what types of waves are involved? These waves cannot be transverse. In order for a transverse wave to propagate, bonding forces must exist among the vibrating particles, and sound is propagated in gases where these forces do not exist. Thus, sound involves a **longitudinal** wave motion. The particles communicate with one another through **vibration** by **shock**, and as a result, the vibration and propagation have the **same line of motion**.

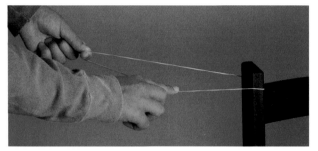
The vibration of a rubber band produces sound waves.

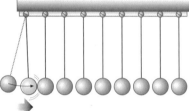

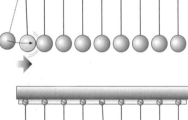

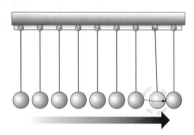

At the left, sound moves like this series of regularly spaced pendulums. The motion of the first is transmitted to the other pendulums by collisions.

HOW THE HUMAN EAR PERCEIVES SOUNDS

4. These vibrations stimulate the auditory organ (Corti's organ) and generate nerve impulses toward the brain, which interprets the sounds.

2. The tiny bones of the middle ear vibrate.

1. The sound waves penetrate into the external auditory canal as far as the tympanic membrane.

3. The vibrations are transmitted through the oval window to the anterior labyrinth (the cochlea or acoustic labyrinth).

WHERE IS SOUND PROPAGATED?

Sound propagates to a greater or lesser extent in **all material media**, independently of their physical state. Since sound is propagated by the collision of particles, it does not propagate in a **vacuum** and it propagates very poorly in **solid, porous** media. As a result, when a soundproof room is desired, its inner surfaces are coated or a double wall is filled with porous materials such as sawdust, cork, or porous plastic. Have you ever watched a movie about the Old West and seen a Native American put his ear to the railroad tracks? Why does he do that? Sound is transmitted much more rapidly in **solid media**, somewhat more slowly in **liquids,** and quite a bit more slowly in **gases**.

Metals are good sound conductors.

THE SPEED OF SOUND

State	Medium	Speed f/s (m/s)
Gases	Oxygen	1,036 (316)
	Air	1,086 (331)
	Nitrogen	1,109 (338)
	Hydrogen	4,133 (1,260)
Liquids	Alcohol	4,182 (1,275)
	Freshwater	4,756 (1,450)
	Salt water	4,920 (1,500)
Solids	Copper	11,808 (3,600)
	Steel	16,400 (5,000)
	Concrete	16,400 (5,000)

PROPERTIES OF SOUND

Some phenomena that are familiar to everyone are the result of the characteristics of waves. Let us examine a couple of them:

- **Reflection:** Sound bounces when it hits a surface and experiences a **sudden change of direction**. One result is an **echo**.

- **Refraction:** If sound passes between two **media** with **different** propagation **velocities**, it **changes course**. At night, the layers of air near the ground are colder than the ones above. As a result, sound that rises toward the sky is diverted back to the ground, and sounds are more audible at night than during the day.

- **Diffraction:** An **obstacle** that has an **opening** becomes a **transmitter** of sound. The sounds are audible even though there is an obstacle between the transmitter and the **receptor**. For example, we can hear what is happening in the street even though we have walked around the corner.

When several reflections of the same sound are heard simultaneously in an area (a hollow sound), we say that there is **reverberation**.

Forces and Their Effects

Motion

Energy

Heat

Fluids

Wave Motion

Sound

Optics

Electricity

Matter

Inside Matter

Mixtures

Pure Substances

Chemical Changes

Alphabetical Subject Index

An echo results when a sound bounces off a large surface.

Since sound moves in all directions, we can hear a sound even though we cannot see the source directly.

INTENSITY OF SOUND

If we touch a gong gently, we displace the metal a bit from its equilibrium position and we hear a **faint** sound. If we give the gong a good rap, the displacement is significant and the sound is **loud.** The magnitude that has varied is the intensity that corresponds to the **amplitude** of the vibratory motion.

Sound displaces a homogeneous medium with uniform rectilinear movement.

A bomb produces a very intense noise (expansive wave) that can carry its energy (destructive power) over a long distance.

TONE

From a distance and without seeing a singer, we can tell if he is a baritone, a bass, or a tenor. We say that the tenor has a **higher** voice than a baritone and that a base has a **deeper** voice. What wave characteristic produces this difference between high and low? The **frequency**, that is, the number of oscillations per second, produces the tone of a sound.

On average, the human ear can hear sounds with frequencies that range from 20 Hz to 20,000 Hz.

In general, children's tone of voice is higher than that of adults, and among adults the tone of women is higher than that of men.

TIMBRE

A flute and an accordion can play the same **note**, but we can tell the difference between them. When an instrument plays a note (a vibration of a certain frequency), the vibration that it produces is not of a single frequency but, rather, a series of **harmonics**. The relative intensity of these harmonics produces the timbre of the sound that helps us to recognize it.

In general, the frequency of a harmonic is a whole multiple of the **basic frequency**.

OPTICS

How do we know that we are spinning around a star that we call the sun? The answer is simple: because we see the star appear every day in the east and vanish in the west. We can see the sun because it sends us light, among other things. This light allows us to see the other terrestrial or celestial objects and makes life possible for green plants. If light is a wave motion, how does it reach the earth through the outer vacuum?

THE NATURE OF LIGHT

Light has the properties of **waves**: it reflects, **refracts, diffracts**, moves in a straight line at a **constant velocity**, transports **energy**, and so forth. Remember: waves need a material medium in order to move. In contrast, light moves through both material media and in a vacuum. With light, the vibrating does not come from a material particle; rather, the waves are **electrical** and **magnetic** in character.

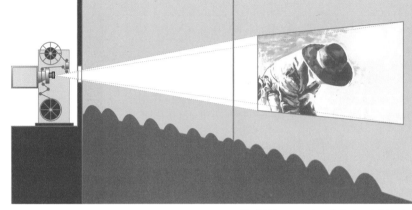

The light from a movie projector has straight edges.

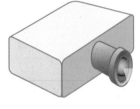

The holes have to be carefully aligned for the ray of light to pass through them since light moves in a straight line.

HOW DOES LIGHT MOVE?

We have already stated that light moves in a **straight line**. To demonstrate that, we have some very simple tests, and some of them occur daily. If we go to the movies, we see that the projector puts out a ray of light that expands to fill the screen, and the **edges** are **straight.** If we place a light source in front of a black screen with a small hole in the middle and then another identical screen, the **holes** have to be lined up for the light to go through the second screen.

A **beam** of light is made up of a tremendous number of light **rays** that leave one point, although in practice we consider a ray to be a very narrow beam.

BEHAVIOR RELATIVE TO LIGHT

The objects that we see are of two types: those that **emit** light and those that **reflect** it partially or totally. The sun is an object in the first category, and we say that it is luminous. The moon, on the other hand, does not emit light but merely reflects the light that comes from the sun. An object is **opaque** when it **absorbs** part of a light beam that hits it and reflects the rest of the beam. It is **transparent** if it allows the light to pass through itself, generally reflecting part of the light and absorbing the rest.

When an object absorbs all the light, we see the color black. When we see white, the object is reflecting all the light.

If an object transmitted all light with no reflection, it would be invisible.

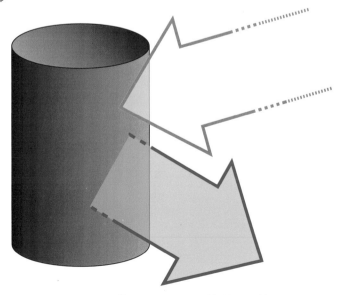

We see a green, opaque object because it has absorbed red light.

The glass in this display window is transparent (we can see through it), but it also reflects since we can see the houses behind us.

HALF-LIGHT (PENUMBRA)

If the light source is large and the obstacle is fairly small (an **extensive light source**), there will be an area of partial shade. What does that mean? Only **some of the light rays** are blocked by the opaque obstacle, resulting in a partially illuminated area. The shape of this area of half-light (penumbra) depends on the size of the obstacle and the distances that exist between the light source, the obstacle, and the screen where the beams of light fall.

For a large obstacle located near the earth's surface, the sun can be considered a pinpoint light source.

SHADOW

We say that a light source is a **pinpoint** when it is very small in comparison with the expanse of light. If a light beam from a pinpoint source hits an opaque object, the part of the beam that hits is blocked while the rest continues on its journey. If we gather the light onto a screen, the object will be reproduced as a black spot with the outline of the object, producing a **shadow**.

When we stand between a light source and a surface, we cast a shadow.

SOLAR ECLIPSES

Sometimes the moon **comes between** the **sun** and the **earth** and produces an area of **shadow** (a total **eclipse**). There is also an area from which part of the sun's surface is visible, and that part provides some illumination to that area of the earth. However, the illumination is only partial (a **partial** eclipse).

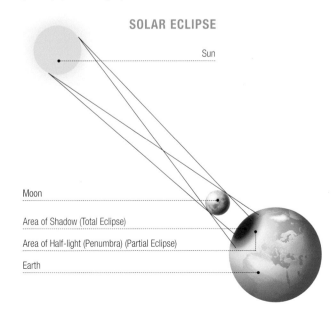

SOLAR ECLIPSE

Sun

Moon

Area of Shadow (Total Eclipse)

Area of Half-light (Penumbra) (Partial Eclipse)

Earth

LUNAR ECLIPSES

Lunar eclipses are a different situation from solar eclipses. In this instance, the **earth** comes between the **sun** and the **moon** to produce an area of shadow on the moon. Since the earth is much larger than the moon, there is no area of penumbra, and the sun acts as if it were a pinpoint light source.

LUNAR ECLIPSE

Total Eclipse of the Moon

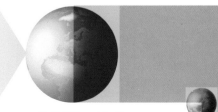

Partial Eclipse of the Moon

Introduction

Forces and Their Effects

Motion

Energy

Heat

Fluids

Wave Motion

Sound

Optics

Electricity

Matter

Inside Matter

Mixtures

Pure Substances

Chemical Changes

Alphabetical Subject Index

MIRRORS

Mirrors can be found in every home, office, and public place. Without these simple instruments, doing many activities would be difficult or impossible. Mirrors can take on various forms depending on the purpose for which they are designed. For example, the rearview mirror of a vehicle does not provide much detail. Of more importance is that it indicates the position of other cars in as large an area as possible.

TYPES OF REFLECTION

As with other types of undulatory phenomena, light is **reflected**. This reflection can be of two types:

- If a beam of light hits a **rough surface**, the reflected rays go off in all directions and no image can be produced with them (**diffuse reflection**).

- If a beam of light hits a smooth surface, the light that leaves the surface obeys the laws of reflection for a beam of light (**specular reflection**).

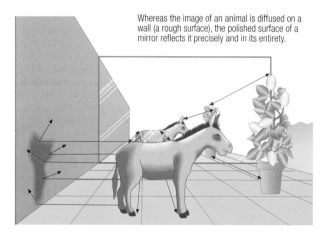

Whereas the image of an animal is diffused on a wall (a rough surface), the polished surface of a mirror reflects it precisely and in its entirety.

Specular reflection makes it possible to see ourselves and do such things as wash ourselves more effectively.

THE LAWS OF REFLECTION

Two laws are observed by every beam of light that falls onto a nonblack, opaque surface:

- The **angle of incidence** equals the **angle of reflection.**

- The **incident,** the **reflected**, and the **normal** rays are all located on the same plane.

The **imaginary straight line** perpendicular to the reflection surface is referred to as the normal; it is used as the basis for measuring angles.

THE LAWS OF REFLECTION

Reflected Ray

Normal

Incident Ray

TYPES OF MIRRORS

In a bathroom mirror, we normally see ourselves in **actual size**, and the surface is flat. In a mirror that is sometimes used for shaving, the face appears very large. If we back up, though, after a moment of confusion, we appear small and in a reversed position. The surface is a **spherical** shape seen from the **inside (concave)**. In automobile rearview mirrors**,** the **images** often appear **small**, but you can see over a large area. Their surface is also curved, but the image is seen from the **outside (convex)**.

TYPES OF MIRRORS

Metallic Part

Concave Mirror

Convex Mirror

Introduction

Forces and
Their Effects

Motion

Energy

Heat

Fluids

Wave
Motion

Sound

Optics

Electricity

Matter

Inside
Matter

Mixtures

Pure
Substances

Chemical
Changes

Alphabetical
Subject Index

 Usually a mirror is a **glass surface** with a **metallic** coating on the back side of the glass. It would be much better if the coating were on the front. These mirrors, though, which are known as **optical mirrors**, are hard to keep in good condition.

VIRTUAL IMAGES

When an image is formed by the prolongation of light rays, it is referred to as a **virtual image**, and it cannot be projected onto a screen.

FLAT MIRRORS

When we place an **object** in front of a flat mirror, we see it as if it were located **behind** the mirror and in **specular symmetry**. This means that the image produced is the same as the object but with left and right reversed. As they are reflected, the light rays reach our eyes and we see the object in the prolongation of these rays.

Flat mirrors are the most common ones. We encounter them in bathrooms and on furnishings, decorations, and so forth.

The arrows indicate the path taken by the light. The light does not reach the image since it is reflected. To the observer, the light seems to emanate from the image, not the object.

How and where an image is seen depends on where it is located. The object is the *R* on top of the optical axis and in front of the mirror.

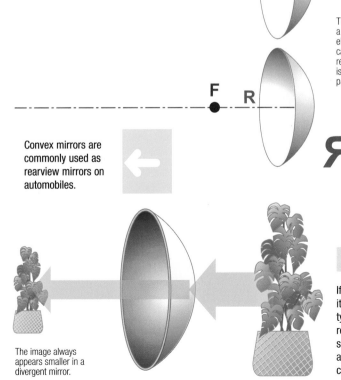

Convex mirrors are commonly used as rearview mirrors on automobiles.

The image always appears smaller in a divergent mirror.

CONCAVE MIRRORS

Imagine a **sphere** of **hollow** glass. If we cut out a spherical cap and plate it with silver on the outside, we have a **concave mirror** if we look at it from the hollow side. These mirrors, which are referred to as **convergent** mirrors, concentrate the light rays, causing them to arrive parallel to the main axis onto a point known as the **focal point**. Concave mirrors form **real images** that can be **projected** onto a screen if the object is farther away than the focal point. The image obtained is large if formed near the focal point or small if formed far away from the focal point. In both cases, the image is **reversed**. If we place the object closer than the focal point, the image is formed upright and large but virtual (i.e., it cannot be projected).

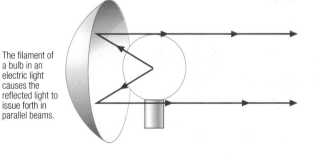

The filament of a bulb in an electric light causes the reflected light to issue forth in parallel beams.

Concave mirrors are commonly used in movie and slide projectors to produce parallel light rays. In some bathrooms are concave mirrors used as an aid in shaving.

CONVEX MIRRORS

If we plate the **inner surface** of a concave mirror with silver and look at it toward the **outward bulge**, we have a **convex** mirror. Mirrors of this type are also called **divergent** mirrors because the light beams that reach their surface and are **parallel** to their main axis are diverted in such a way that they **separate**, but their **extensions** come back together again in a point known as the **focal point**. The images produced by a convex mirror are always **virtual**, **upright**, and **small**.

LENSES

If we look around us, we see that many of our friends and acquaintances wear glasses in order to see clearly. Upon further consideration, we might even say that all of us peer through lenses. Why? The crystalline lens is part of the eye of humans and animals and is made up of a convex lens that concentrates light into a point. In addition, all devices referred to as *optical* commonly contain some type of lens for retaining or projecting images.

It appears to the observer that the pencil is broken at its point of contact with the surface of the water due to the deviation of light rays.

A diamond accumulates light like a barrel into which we funnel more water than comes out from the spigot.

A **diamond** is so shiny because light enters at over 186,000 miles (300,000 km) a second and flows inside it at only 76,800 miles (124,000 km) per second; so **light is retained** inside the diamond.

REFRACTION

When light **hits a surface at an angle** and that surface separates two media through which light's **propagation velocity** is different, two things can happen:

- If the **velocity** in the first medium is **greater** than in the second one, the angle that the incident beam forms with the normal is **greater** than the one formed by the normal with the refracted beam (it comes **closer**).

- If the **velocity** in the first medium is **less** than in the second one, the angle that the incident beam forms with the normal is **smaller** than the one formed by the normal with the refracted beam (it moves **farther away**).

Just as with reflection, the normal is the imaginary perpendicular line between the plane that separates the media at the point of incidence.

REFRACTION INDEX

The refraction index in a medium is the quotient between the speed of light in a vacuum and the speed of light in that medium.

Substance	Vacuum	Air	Water	Paraffin	Glass	Diamond
Refraction Index	1	–1	1.33	1.44	1.52	2.42

WHAT IS A LENS?

A lens is a **transparent body** defined by two **smooth** surfaces of which at least one is **spherical**. We must keep in mind that within this body, light has to propagate at a different speed (generally lower) than in the outside. Depending on the **shape** of these surfaces, we can identify lenses of different types and with **different behaviors.**

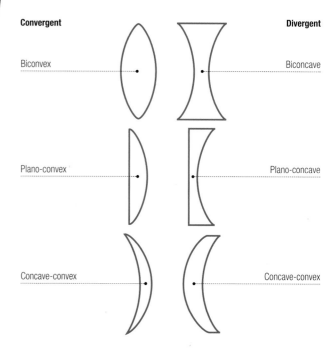

Convergent / Divergent

Biconvex / Biconcave

Plano-convex / Plano-concave

Concave-convex / Concave-convex

TYPES OF LENSES

Lenses may be **convergent** or **divergent** depending on whether they **join** the light rays that go through them or **separate** the rays. If the lens is made of glass or plastic, which is common, and the outside medium is air, convergent lenses are **thicker in the center** and thinner at the edges, and divergent lenses are **thicker at the edges** than in the center. Thus, depending on the shape of the separation surfaces, lenses can be **convergent** (biconvex, plano-convex, or concave-convex) or **divergent** (biconcave, plano-concave, or concave-convex).

Introduction

Forces and
Their Effects

Motion

Energy

Heat

Fluids

Wave
Motion

Sound

Optics

Electricity

Matter

Inside
Matter

Mixtures

Pure
Substances

Chemical
Changes

Alphabetical
Subject Index

ELEMENTS OF A LENS

If the outside medium were denser than that of the lens, the shape of the convergent and divergent lenses would be reversed.

- **Optical axis** or **principal axis**: an imaginary line that goes through the lens perpendicularly at its thinnest (divergent) or thickest (convergent) point.

- **Optical center**: the point of intersection between the optical axis and the lens.

- **Focus**: the point where the light beams parallel to the optical axis (convergent) or its extension (divergent) gather.

- **Focal distance**: the distance in meters between the optical center and the focus. The focal distance is either positive (convergent) or negative (divergent).

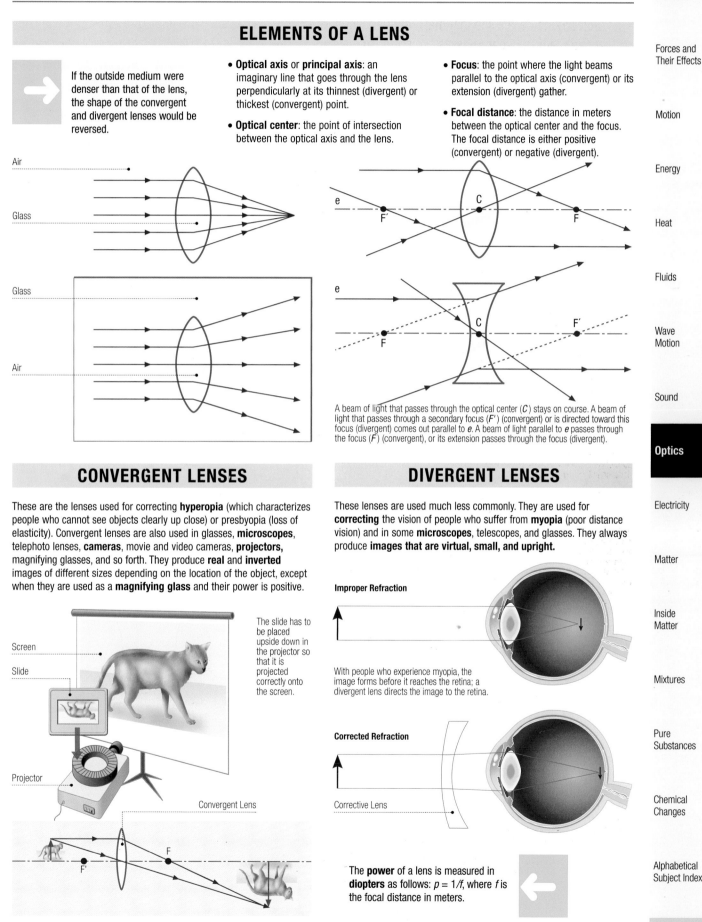

A beam of light that passes through the optical center (*C*) stays on course. A beam of light that passes through a secondary focus (*F'*) (convergent) or is directed toward this focus (divergent) comes out parallel to *e*. A beam of light parallel to *e* passes through the focus (*F*) (convergent), or its extension passes through the focus (divergent).

CONVERGENT LENSES

These are the lenses used for correcting **hyperopia** (which characterizes people who cannot see objects clearly up close) or presbyopia (loss of elasticity). Convergent lenses are also used in glasses, **microscopes**, telephoto lenses, **cameras**, movie and video cameras, **projectors**, magnifying glasses, and so forth. They produce **real** and **inverted** images of different sizes depending on the location of the object, except when they are used as a **magnifying glass** and their power is positive.

The slide has to be placed upside down in the projector so that it is projected correctly onto the screen.

DIVERGENT LENSES

These lenses are used much less commonly. They are used for **correcting** the vision of people who suffer from **myopia** (poor distance vision) and in some **microscopes**, telescopes, and glasses. They always produce **images that are virtual, small, and upright.**

Improper Refraction

With people who experience myopia, the image forms before it reaches the retina; a divergent lens directs the image to the retina.

Corrected Refraction

The **power** of a lens is measured in **diopters** as follows: $p = 1/f$, where *f* is the focal distance in meters.

ELECTROSTATIC PHENOMENA

Is it true that electricity is a fairly recent invention? There is a belief that electricity was invented in the nineteenth century, but nothing is further from the truth. In the first place, electricity is not an invention; it is a natural phenomenon that has existed on earth ever since the remotest times. The Greeks of the classical age were the first ones to notice the phenomenon that they called electric when they saw that amber attracted small, light objects such as bird feathers when it was rubbed with animal fur.

TYPES OF CHARGES

We enjoy a higher standard of living thanks to electricity.

Charges acquired by a conductor, such as all metals, are more difficult to detect because they are less concentrated than in insulators.

Matter can behave in two different ways when it is rubbed. The charges remain localized in a single point (**insulators**), or they spread out over the entire surface (**conductors**).

The most ancient way to put an electrical charge into an object is to **rub** it with something else. If we rub two glass rods with a woolen cloth and suspend them by threads, we see that the two rods **repel one another**. However, if we rub one glass rod and another one made of plastic, when they are hung up, forces of **attraction** appear. This indicates the existence of **two different types of charge**: the type acquired by glass when it is rubbed with wool and another that is acquired by plastic. The first type of charge is referred to as **positive**, and the second is said to be **negative**.

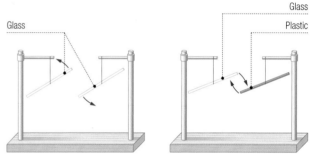

The two glass rods tend to move apart, but the glass rod and the plastic rod move closer together.

An object is said to be neutral when it contains the **same number** of **positive** and **negative** charges. In **conductors,** these charges are capable of moving.

COULOMB'S LAW

When two electrically charged objects are near one another, a force appears between them known as an **electrostatic force**. It is directly proportional to the **value of the charges** and inversely proportional to the square of the **distance that separates them**. These can be forces of **attraction** or **repulsion**, depending on whether the two charges are of a different or the same type. The mathematical expression for Coulomb's law is:
$F = K \dfrac{q \cdot q'}{d^2}$, where K is a constant dependent on the medium.

The value of K in a vacuum or in air is 9×10^9 N · m²/c², that is, the force that appears between two charges of 1 C located at a distance of 1 meter is 9,000,000,000 N.

Wool

Glass

Positive Charges

Iron

Wool

Wooden Handle

Positive Charges

WAYS TO CHARGE AN OBJECT

We have already seen that an object can become electrically charged by rubbing. However, there are other ways to accomplish the same thing: **contact** and **induction**.

- **Contact:** If an electrically charged object comes into contact with another that is initially neutral, it **communicates its charge** to the neutral one.

- **Induction:** If an electrically charged object is located **near** a **conductor**, the charges of the conductor move in such a way that it **polarizes**. If the conductor is discharged by touching one of the poles with a finger, the object remains charged with the **sign opposite** that of the pole that was touched.

In touching the negative rod, the entire metallic ball of the electroscope is charged negatively. Thus, the aluminum strips repel one another and move apart.

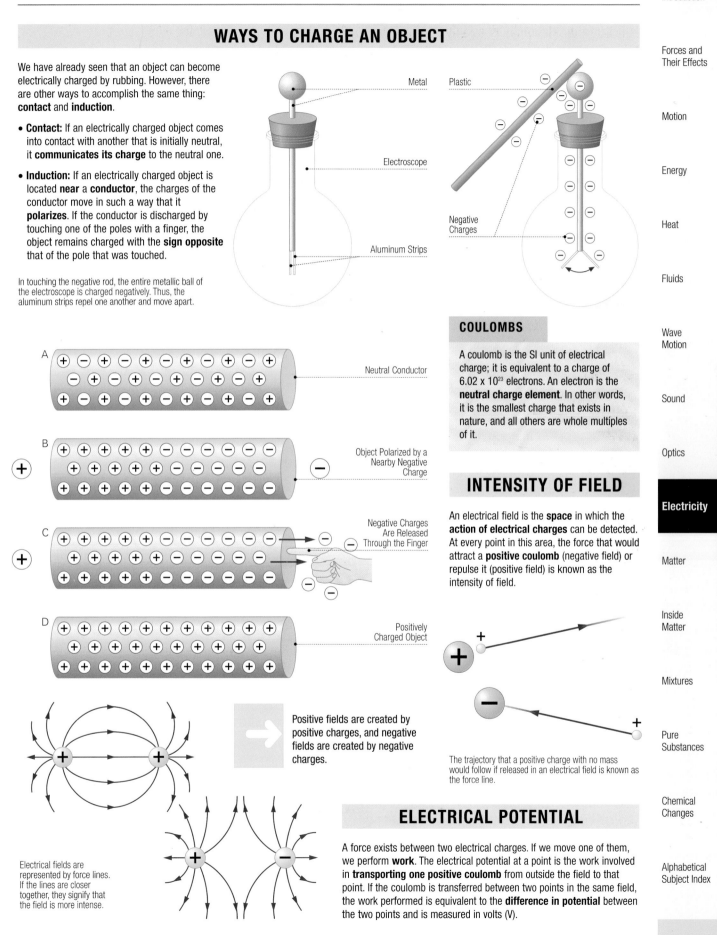

Metal

Plastic

Electroscope

Negative Charges

Aluminum Strips

A — Neutral Conductor

B — Object Polarized by a Nearby Negative Charge

C — Negative Charges Are Released Through the Finger

D — Positively Charged Object

Electrical fields are represented by force lines. If the lines are closer together, they signify that the field is more intense.

Positive fields are created by positive charges, and negative fields are created by negative charges.

COULOMBS

A coulomb is the SI unit of electrical charge; it is equivalent to a charge of 6.02×10^{23} electrons. An electron is the **neutral charge element**. In other words, it is the smallest charge that exists in nature, and all others are whole multiples of it.

INTENSITY OF FIELD

An electrical field is the **space** in which the **action of electrical charges** can be detected. At every point in this area, the force that would attract a **positive coulomb** (negative field) or repulse it (positive field) is known as the intensity of field.

The trajectory that a positive charge with no mass would follow if released in an electrical field is known as the force line.

ELECTRICAL POTENTIAL

A force exists between two electrical charges. If we move one of them, we perform **work**. The electrical potential at a point is the work involved in **transporting one positive coulomb** from outside the field to that point. If the coulomb is transferred between two points in the same field, the work performed is equivalent to the **difference in potential** between the two points and is measured in volts (V).

Introduction

Forces and Their Effects

Motion

Energy

Heat

Fluids

Wave Motion

Sound

Optics

Electricity

Matter

Inside Matter

Mixtures

Pure Substances

Chemical Changes

Alphabetical Subject Index

ELECTRICAL CURRENT

What is it about electrical current that makes it our favorite among all types of energy? There are many answers to this question. First of all, we have to keep in mind that electrical energy originates from other energy sources that may or may not be clean.

Electricity has the highly esteemed property of being distributable continually and fairly cheaply over long distances. In addition, the sources, which are sometimes disagreeable, can be located far away from urban population centers.

WHAT IS CURRENT?

Electrical current is the **passage of electrical charges** through a conductor between whose ends, known as **terminals**, there exists a **difference in potential**. The conductors may be solid, liquid (generally solutions), or gaseous.

- In **solid** conductors, the moving charges are negative **electrons**.

- **Positive** and negative charges move in **solutions**; they circulate in opposite directions.

- **Gases** are **poor electrical conductors**; but if given a certain difference in potential, they can turn into conductors.

Gaseous Conductor

Solid Conductor

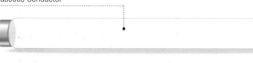

A fluorescent bulb contains a gaseous electrical conductor; an electrical cord, on the other hand, is a solid conductor.

Electrical current is transported from its production point through high-tension wires.

WHAT CREATES THE DIFFERENCE IN POTENTIAL?

The difference in potential is created by machines known as **current generators** that consume various types of energy and transform them into electrical energy. Current generators include **batteries** that transform chemical energy into electrical energy;

dynamos or **generators** that some bicycles have for illuminating the road and that transform mechanical energy into electrical current; and **alternators**, like the ones on automobiles, to provide energy for the horn, lights, starter motor, and other applications.

Generator

Battery

TYPES OF CURRENT

There are basically two types of current: **direct current** and **alternating current**. In the first instance, the electrical charges always go in the **same direction.** With alternating current, the charges have a **back-and-forth** motion. The negative charges move from the lowest potential to the highest, and the positive ones (in solutions and in gases) move from the greatest to the smallest. With direct current, which is created by batteries and generators, the terminals do not change polarity, so the **same one** always has the **highest potential**. With **alternating current**, however, which is created by alternators, the **highest potential** continually **changes** terminal. Alternating current is the one generally used for **long-distance transport** and for residences.

The wheel of a bicycle generator turns through friction against the tire. Nowadays generators are commonly replaced by batteries that activate a headlight and a taillight.

Introduction

Forces and
Their Effects

Motion

Energy

Heat

Fluids

Wave
Motion

Sound

Optics

Electricity

Matter

Inside
Matter

Mixtures

Pure
Substances

Chemical
Changes

Alphabetical
Subject Index

In commercial alternating current, the polarity changes 50 to 60 times a second.

There are **rectifiers** that change alternating current to direct, which is the type of current used in electronic devices.

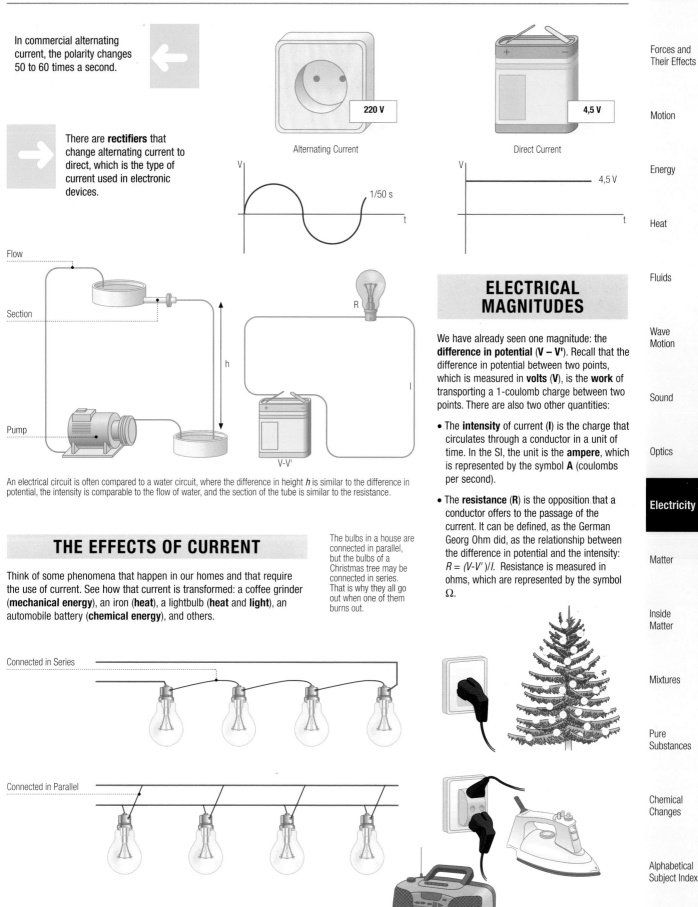

220 V

Alternating Current

1/50 s

4,5 V

Direct Current

4,5 V

Flow

Section

h

Pump

R

I

V-V'

An electrical circuit is often compared to a water circuit, where the difference in height *h* is similar to the difference in potential, the intensity is comparable to the flow of water, and the section of the tube is similar to the resistance.

ELECTRICAL MAGNITUDES

We have already seen one magnitude: the **difference in potential** (**V − V'**). Recall that the difference in potential between two points, which is measured in **volts** (**V**), is the **work** of transporting a 1-coulomb charge between two points. There are also two other quantities:

- The **intensity** of current (**I**) is the charge that circulates through a conductor in a unit of time. In the SI, the unit is the **ampere**, which is represented by the symbol **A** (coulombs per second).

- The **resistance** (**R**) is the opposition that a conductor offers to the passage of the current. It can be defined, as the German Georg Ohm did, as the relationship between the difference in potential and the intensity: $R = (V - V')/I$. Resistance is measured in ohms, which are represented by the symbol Ω.

THE EFFECTS OF CURRENT

Think of some phenomena that happen in our homes and that require the use of current. See how that current is transformed: a coffee grinder (**mechanical energy**), an iron (**heat**), a lightbulb (**heat** and **light**), an automobile battery (**chemical energy**), and others.

The bulbs in a house are connected in parallel, but the bulbs of a Christmas tree may be connected in series. That is why they all go out when one of them burns out.

Connected in Series

Connected in Parallel

MAGNETISM AND ELECTROMAGNETISM

How many magnets and electromagnets are in a typical house? A few? Dozens? Maybe more? Consider for a moment the compass we use when we go hiking, the magnets we have on the refrigerators for shopping lists and other reminders, the closures on many cabinet doors in the kitchen and surely other rooms, all the electrical devices that use a motor—blenders, coffee grinders, garbage disposals, vacuum cleaners—and so forth.

MAGNETS

In nature, there are certain minerals, such as **magnetite**, that have the characteristic of exerting a fairly strong **attraction** to **iron objects** and, to a lesser degree, to objects containing **nickel** and **cobalt**. Materials can be created **artificially** that possess this same characteristic in a **temporary** or **permanent** way (soft iron and steel, respectively). The force of attraction is stronger at the ends of a magnet. They are known as the **north** and **south poles** (negative and positive, respectively) because a magnet that is free to turn horizontally aligns itself to point north and south. The earth thus acts like a magnet whose negative pole is located near the geographic North Pole.

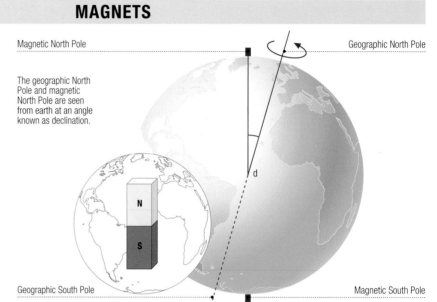

Magnetic North Pole · Geographic North Pole

The geographic North Pole and magnetic North Pole are seen from earth at an angle known as declination.

d

N
S

Geographic South Pole · Magnetic South Pole

 The rare-earth metal known as **neodymium** can be used to create much stronger magnets than ones made of iron.

If a **magnet** is cut in two in the middle, the two poles are not separated from one another. Rather, two smaller magnets are created.

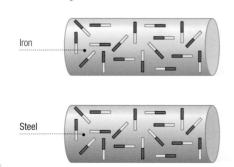

The motor of a household vacuum cleaner functions with electromagnets.

MAGNETIC INDUCTION

When a piece of iron or steel is **touched with a magnet**, the piece **takes on magnetic properties**. In the case of iron, those properties vanish. With steel, they remain for a certain time. A weak magnet can be made by **rubbing** a bar of iron in the same direction **with a magnet**. We can imagine magnetic materials as **small, disorganized magnets** or **domains** whose effects cancel one another out. When we touch them with a magnet, the magnet's positive pole attracts the negative poles of the materials and vice versa; that is how small domains **orient themselves**. When contact with the magnet ends, iron becomes disorganized once again, but steel does not.

Iron returns to its original state when the magnet is taken away; however, steel remains magnetized for a certain time.

Iron

Steel

CURRENT AND MAGNETISM

In the eighteenth century, a Dane named Christian Oersted performed the following experiment. He placed a compass (a magnet that turns freely) onto a straight **conducting wire**. The compass did not **point** north and south but, rather, **perpendicular to the wire**. This phenomenon indicated that electrical current produced magnetic effects.

Introduction

Forces and Their Effects

Motion

Energy

Heat

Fluids

Wave Motion

Sound

Optics

Electricity

Matter

Inside Matter

Mixtures

Pure Substances

Chemical Changes

Alphabetical Subject Index

OERSTED'S EXPERIMENT

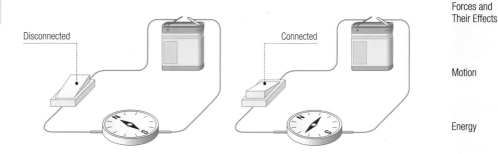

Disconnected

Connected

MAGNETIC FIELD

A magnetic field is the area affected by a magnet or an electromagnet. Magnetic fields are represented by lines of force.

THE LAW OF THE RIGHT HAND

If a conducting wire goes through a surface at a right angle, it creates a magnetic field whose force lines are **concentric with the wire**. The direction of the field is the one indicated by the **fingers** of our right hand holding the wire, and the **thumb** points in the **line of motion of the current**. This rule allows us to determine the field created by a spiral and, by extension, of many spirals, which are known as a **solenoid**.

THE LAW OF THE RIGHT HAND

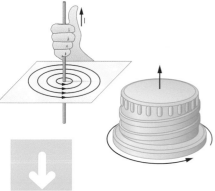

Just as an electrical current creates a **magnetic field**, a variable magnetic field can **create** an **electrical current**.

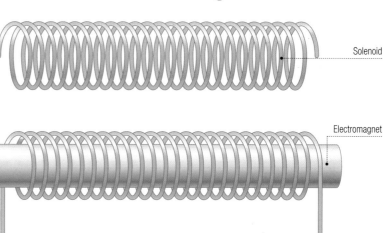

Spiral

Solenoid

Electromagnet

If an iron core is added to a solenoid, it is called an **electromagnet**, and the intensity of the magnetic field it produces increases significantly.

APPLICATIONS

Magnets and electromagnets have many applications, including the following:

- **Loading magnets**: These are large electromagnets that are used to move iron materials.

- **Relays**: These are switches activated by an electromagnet in order to avoid, to the extent possible, sparks from contact; they can be activated from a distance.

- **Measuring devices**: Many ammeters, voltmeters, and so forth are based on the force produced by an electromagnet.

- **Motors and current generators** consist of electromagnets or movable magnets.

Large electromagnets are used in scrap yards to move iron materials.

THE CLASSIFICATION OF MATTER

Everything around us is made up of matter. Matter is the air that we breathe, the book we are reading, the water that we drink, and so forth. Matter has certain **characteristics** that help us distinguish one type from another.

A single property does not indicate what substance or substances we are dealing with. However, the sum of several properties may, just as saying that a person is blonde does not indicate a specific person. In order to specify precisely, we have to indicate some of the person's other characteristics.

VISUAL CLASSIFICATION

If we look at any of the things that surround us, we see materials of two different types:

- **Heterogeneous:** At a simple glance or with the aid of a magnifying glass, we can see **parts** that are **differentiated** by color, physical state, and so forth. The physical or chemical **properties** that we identify **depend** on the **sample selected** from the whole mass.

- **Homogeneous:** We cannot detect any **differentiated parts** even with the aid of a magnifying glass. The physical and chemical **properties are the same** at every point in the entire mass.

The cloudy water of a river, like the water in this glass, is an example of a heterogeneous mixture. The densest materials are the first ones to settle to the bottom.

HETEROGENEOUS MIXTURES

Heterogeneous materials, which are also referred to as **heterogeneous mixtures**, can be separated into their components by purely **mechanical means**, such as **filtration**, **decanting**, **centrifuging**, **flotation**, **extraction**, and so forth. This does not mean that there are no more energetic ways to achieve separation. These mechanical methods are commonly preceded by preparatory operations such as **grinding** or **crushing** in the case of solid materials.

The test tubes spin around rapidly, and the solids in the mixture settle out and become compacted. Separating them from the liquid by decanting then becomes easy.

Every one of the differentiated parts of a heterogeneous mixture is known as a **phase**, and each phase may be homogeneous.

In decanting, the mixture is allowed to stand until the solids settle out completely, and then they are separated by pouring off the liquid.

Filtration serves to separate solids from liquids in heterogeneous mixtures.

Introduction

Forces and
Their Effects

Motion

Energy

Heat

Fluids

Wave
Motion

Sound

Optics

Electricity

Matter

Inside
Matter

Mixtures

Pure
Substances

Chemical
Changes

Alphabetical
Subject Index

HOMOGENEOUS MATERIALS

Just because an examination or a study of properties indicates homogeneity, that does not mean that we are dealing with a single substance. Homogeneous matter can be of two types:

- **Solutions**: These are homogeneous **mixtures** made up of a **solvent** and one or more dissolved substances known as **solutes**. The components of a solution can be separated using **changes of state** such as distillation and freezing.

- **Pure substances:** These are substances that do not separate into different parts through changes of state and that have **fixed, specific properties.**

Icebergs are composed of ice from freshwater even though they were formed from salty seawater.

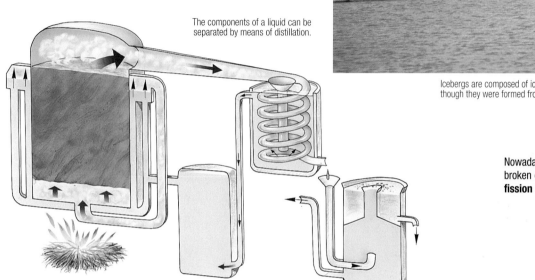

The components of a liquid can be separated by means of distillation.

Nowadays elements can be broken down through **nuclear fission** reactions.

PURE SUBSTANCES

Every pure substance has a set of **physical** and **chemical properties** that define it. The physical properties include melting point, boiling point, color, hardness, density, and others. The chemical properties include stability plus **reactivity** with water, oxygen, acids, and bases, and so forth. Pure substances may be of two classes:

- **Compounds**: These are substances that can form by reaction between **simpler pure substances** or that can **break down** by action of heat or electricity.

- **Elements or simple substances**: These are substances that **cannot break down** by action of heat or electricity **nor form** from simpler substances.

Electric Oven

Carbon Dioxide

Calcium Hydroxide

Marble

White marble is a pure substance, but we know that it is not an element because when it is heated, it breaks down into calcium hydroxide and carbon dioxide (gas).

To date, there are 113 natural and synthetic elements, but most materials can be formed using only about 20 of them.

SYMBOLS AND FORMULAS

The **elements** are represented using **symbols**. The symbols commonly use the **capitalized first letter** of the **Latin** name of the element. If another element uses that first letter, then the second letter, in lowercase, is added. For instance, N is nitrogen but Na is sodium (*natrium* in Latin).

Compounds are represented by **formulas**. The formula contains the **symbols of the elements** that make up the compound with a **subscript number** that indicates the **proportion** in which each element is included.

KINETIC THEORY

If we were asked, "What is iron?" our response would almost certainly be something like the following: Iron is a solid, gray metal that conducts. . . . This answer is correct only up to a certain point since it is a bit vague. How so? In indicating the state of the aggregate of iron or any other substance, we have to specify the conditions. So in this case, we should say it is solid at ambient temperature, for the state would certainly be different at other temperatures.

KINETIC THEORY

This theory holds that every particle of matter, as long as it is not at absolute zero (−459.67°F/−273.15°C), is in **constant motion** and that its **kinetic energy** is directly related to **temperature.** Thus, when we heat an object, it stores the energy received in the form of kinetic energy (an increase in temperature) and **potential energy** (expansion). The speed of all particles is not the same. For every temperature, though, is a corresponding **average speed**.

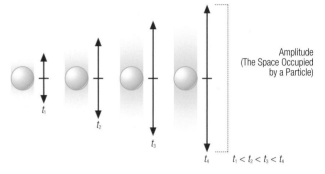

Amplitude
(The Space Occupied by a Particle)

$t_1 < t_2 < t_3 < t_4$

As the amplitude of vibration increases, the space occupied by each particle increases.

SOLIDS

According to the kinetic theory, the particles of solids can move in just one way: **vibration**. Each particle vibrates around an equilibrium point at a **frequency** that depends on its mass and the **forces of interaction** with neighboring particles. Earlier when we said that the solid state is a **rigid state**, we were defining a **macroscopic property** (which can be seen). However, that did not correspond to the **intimate nature** of solids since they are in constant motion and always changing shape even though they appear rigid on the outside.

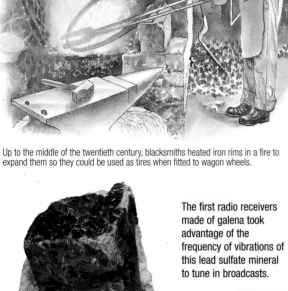

Up to the middle of the twentieth century, blacksmiths heated iron rims in a fire to expand them so they could be used as tires when fitted to wagon wheels.

The first radio receivers made of galena took advantage of the frequency of vibrations of this lead sulfate mineral to tune in broadcasts.

The appearance of galena, the mineral used as the first semiconductor to detect radio waves.

The particles of solids behave as if they were joined together by springs.

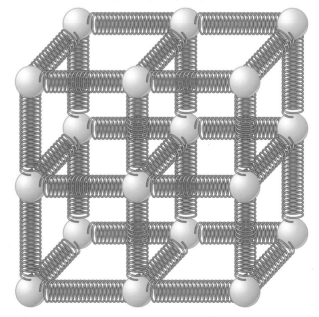

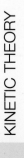

Naphthalene tablets evaporate in a couple of weeks.

DO SOLIDS EVAPORATE?

Surely we have all seen balls, disks, or packets of a white, crystalline, and foul-smelling solid substance placed into closets where wool clothing is stored. This solid is usually **naphthalene** or mothballs (**paradichlorobenzene**); it is used to prevent damage caused by **moths**. After a while, this solid disappears by evaporation. That is how its disagreeable odor spreads throughout the room.

LIQUIDS

The particles of liquids can move with greater freedom than those of solids. Particles in liquids can move **in all directions** and cover great distances with just one limitation: they must always remain **in constant contact** with one another. This contact is broken only when one of the particles reaches the surface with **enough speed** and is traveling in the **right direction** to escape into the air. In such cases, we say that the liquid is **evaporating**.

GASES

Gas particles have **total freedom** of motion. A particle can occupy **any point in its container** that is not already occupied by another particle, and the **forces of interaction** among the particles are practically **nonexistent**. The gas laws we have already seen were calculated as if these forces were nil, so they are known by the name of **ideal gas** laws.

IDEAL GASES

A gas behaves like an ideal gas when it is far away from conditions of condensation, in other words, at **high temperature** and **low pressure.**

LAWS AND MOTION

In studying the laws of gases, we saw that when **temperature increases** at a constant volume, **pressure increases**. Why? As the temperature rises, the speed of the particles increases, and the **violence** of the collisions among them and with the walls of the container thereby increases. The pressure is the force of these collisions divided by the surface area. We also saw that **pressure can increase by decreasing the volume** at a constant temperature. The explanation of kinetic theory is easy: as the volume decreases without lowering the temperature, the **number of collisions** increases without decreasing their intensity. As a result, the pressure goes up.

Particles that land at an excessively oblique angle do not evaporate even though they have enough speed; they bounce on the top like a rock skipped across the surface of a lake.

If the speed of carnival bumper cars were increased, they would become dangerous because the number and intensity of collisions would increase.

Introduction

Forces and Their Effects

Motion

Energy

Heat

Fluids

Wave Motion

Sound

Optics

Electricity

Matter

Inside Matter

Mixtures

Pure Substances

Chemical Changes

Alphabetical Subject Index

WHAT IS A CHANGE OF STATE?

A change of state is the struggle between two phenomena that have **opposing effects**: **motion** and the **forces of interaction**.

- Motion tends to separate the particles. Its effect **increases with temperature**, which increases speed.

- The forces of interaction are the sum of all the electrical and gravitational forces that tend to **join the particles**. These forces vary only slightly with temperature but are **sensitive to pressure**.

VAPOR PRESSURE

At the beginning of this section, we saw that both solids and liquids can evaporate. Evaporation is the passage of particles from solids or liquids into the air. If evaporation occurs inside a **closed container**, as the number of gas particles increases, **the pressure increases**. For every substance at a given temperature, there is a pressure that is reached and increases no further. So does evaporation cease? Evaporation continues, but the particles in the air return to their initial state, so we do not perceive any change. The maximum pressure of each substance is known as its **vapor pressure**.

When the particles that make up the ice cube contact the hot liquid, they lose their condition of equilibrium.

The passage from the liquid to the solid state is called **solidification**, and it occurs at the same temperature as does melting.

If the vapor pressure of the solid is higher than the **outside pressure** (generally the atmospheric pressure), the solid changes directly to gas. This phenomenon is known as progressive **sublimation**.

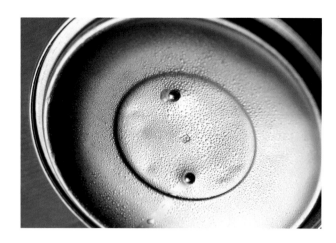

Vapor pressure increases with temperature in both solids and liquids.

When you take the cover off a pan containing water, even though it is cold, drops of water are on the cover from the condensed vapor.

FUSION

Is there anyone who has never watched an ice cube melt in a glass? This phenomenon, evidently common and simple, is more complex than it seems. When the ice reaches a certain **temperature**, the **amplitude of the vibrations** of its particles is so great that the forces of interaction are not capable of returning each particle to its equilibrium position. Each particle is therefore **free**, but its velocity is not adequate to leave its companions, so it passes into the liquid state.

With the first warm days, the snow melts and forms little rivulets that join one another and eventually make up a larger river.

BOILING

At a certain temperature when **vapor pressure** equals the **pressure outside** the container, usually the atmospheric pressure, **bubbles** form inside the liquid and rise to the surface of the liquid, where they pop violently. This phenomenon is known as **boiling**. At this temperature, the speed of the particles is so great that they completely overcome the forces of interaction and become entirely free.

The return of a gas to the liquid state is referred to as **condensation** or **liquefaction**, and it occurs at the same temperature as boiling.

LAWS OF CHANGES OF STATE

All changes of state are governed by the same laws:

- At a given pressure, every substance has a characteristic **melting point** and **boiling** or **sublimation point**.

- The **temperature** remains **constant** as long as the change of state continues.

- Melting, boiling, or progressive sublimation takes place by means of **absorption** of a certain quantity of heat that is characteristic of each substance. In condensation, solidification, or regressive sublimation, this **same quantity of heat is set free.**

Vacuum Pump

If the pressure is sufficiently low, water can be made to boil with the heat of a person's hands.

We can tell that water has reached its boiling point when bubbles form that release water vapor when they burst.

S = Solid
L = Liquid
G = Gas

Temperature

Boiling

Melting

Time

During melting, the solid and liquid phases coexist; during boiling, the liquid and gaseous states coexist.

Forces and Their Effects

Motion

Energy

Heat

Fluids

Wave Motion

Sound

Optics

Electricity

Matter

Inside Matter

Mixtures

Pure Substances

Chemical Changes

Alphabetical Subject Index

EVAPORATION AND BOILING

You may have noticed that both evaporation and boiling are the passage from the liquid to the gaseous state. It is worthwhile to take note of the differences between these two. **Evaporation** takes place at **any temperature**, although it occurs more quickly as the temperature increases. **Boiling** occurs at a **specific temperature** for every substance. Evaporation takes place only at the surface. So in order to encourage evaporation, the liquid can be spread out, as in washing the floor. Boiling takes place at every point in the liquid.

Amorphous solids such as wax and glass have no melting temperature since they **gradually soften**; however, they do have a boiling point. That is why they are considered to be **liquids of high viscosity**.

ATOMS

Would it make sense for me to describe what your house is like? Of course not. You can see it, so you know if perfectly. If I speak with a friend and tell stories about a visit to your house, my friend may imagine what it is like and form an accurate picture, but there is no guarantee of that. As the conversation goes on, the shape and contents of the imaginary house change and come closer to reality. Like this example, we cannot see inside matter, so we have to imagine what it is like based on experience, by creating a model.

Even in the darkness, we can use our sense of touch to produce a general idea of what is in a room.

Not all gaseous elements are diatomic. The **noble gases** are made up of a single atom. Some vapors, such as those of **phosphorous** and **sulfur**, are made up, respectively, of four (P_4) and eight (S_8) atoms.

We can imagine atomic solids and liquids as different little balls for every element, gaseous elements as pairs of equal balls, and compounds as clusters of different balls.

Atomic Elements	Molecular Elements	Compounds
Potassium	Oxygen	Water
Strontium	Hydrogen	Hydrogen Chloride
Barium	Chlorine	Chlorous Acid

DALTON'S MODEL

Dalton imagined what matter is like by studying the **relationships** between the **masses** of the reagents and the products of a chemical reaction. His idea of matter was very close to the one we presently have; it can be summarized in three points:

- The **elements** are made up of **atoms**, and all the atoms of a single element are the same.

Dalton was not aware of the existence of **isotopes**: atoms of a single element with at least one property that is different, such as mass.

- **Compounds** are made up of **molecules** or **ions**. All molecules of a particular compound are the same. Molecules are **clusters of atoms**.

- Molecules are broken down in **chemical reactions**, and the resulting **atoms** are **rearranged** in a different way.

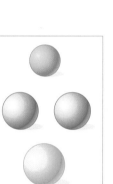

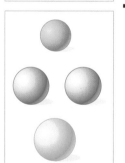

The initial compound breaks apart and gives rise to two different compounds.

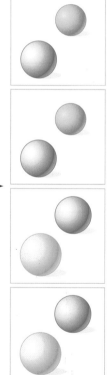

GASES

Avogadro studied the relationships among the **volumes** of gases in a reaction and developed a model to explain the behavior of gases. He said the same **number of particles** of any gas occupy the **same volume** as long as measurements are taken under the same conditions of pressure and temperature. There were a few contradictions between Dalton's model of matter and Avogadro's gas model. Those contradictions were resolved by adding to the first point of Dalton's model that the **gaseous elements** are made up of **diatomic molecules** (two identical atoms).

WHAT ARE ATOMS LIKE?

No one has seen an atom, but throughout the twentieth century there were several models of atoms.

- For **Thompson**, the atom was a **positive mass** with imbedded electrons.

- For **Rutherford**, the atom had a very tiny **nucleus** with positive and neutral masses and a **circular corona** of negative **electrons**. The radius of the electron corona was around ten thousand times that of the nucleus.

- For **Bohr**, the atom was like a solar system. There is a nucleus like the one imagined by Rutherford, but the electrons are arranged in various **levels** with different energies and the shape of the orbits may be circular or elliptical.

- The **present** atom model is much more complex. Suffice it to say that it no longer speaks of orbits and introduces the concept of an **orbital**. An orbital is an **area** of space where the greatest **probability** exists that there is an electron.

MODELS OF ATOMS

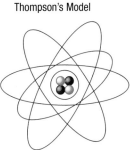

Thompson's Model

Rutherford's Model

Bohr's Model

Schrödiger's Model (Current)

MOLECULAR MASS

The mass of a molecule is measured by adding together the atomic masses of the elements that make it up.

The mass of a sodium atom is equivalent to that of 23 hydrogen atoms. Thus, its atomic mass is 23.

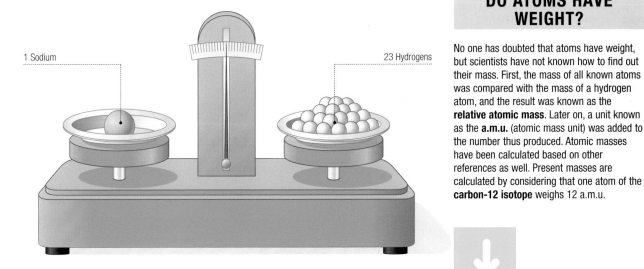

1 Sodium

23 Hydrogens

DO ATOMS HAVE WEIGHT?

No one has doubted that atoms have weight, but scientists have not known how to find out their mass. First, the mass of all known atoms was compared with the mass of a hydrogen atom, and the result was known as the **relative atomic mass**. Later on, a unit known as the **a.m.u.** (atomic mass unit) was added to the number thus produced. Atomic masses have been calculated based on other references as well. Present masses are calculated by considering that one atom of the **carbon-12 isotope** weighs 12 a.m.u.

THE MOLE

In the laboratory, chemists cannot measure in a.m.u. The relative atomic mass **in grams** is called the **atom gram** and the quantity of matter it represents is the mole. The **mole** is the unit in the SI system for the quantity of a substance. One mole of any substance always has the same number of particles as one mole of another substance.

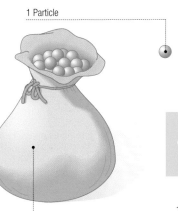

1 Particle

1 Mole:
602,000,000,000,000,000,000,000 Particles

Avogadro's number is the number of particles contained in a mole: 6.02×10^{23}, in other words, 602,000,000,000,000,000,000,000 particles.

In a mole of water (18 g) there are 6.02×10^{23} molecules, millions of times more than all the hairs on the heads of all humans in the last 20 centuries added together.

Introduction

Forces and Their Effects

Motion

Energy

Heat

Fluids

Wave Motion

Sound

Optics

Electricity

Matter

Inside Matter

Mixtures

Pure Substances

Chemical Changes

Alphabetical Subject Index

PERIODIC CLASSIFICATION

When does it become necessary to have a classification system? Imagine that you have a bookshelf with only four books. It would be absurd to waste much time thinking about how to arrange them. You can find the book you want just by looking. Now suppose that you have 120 books in your room.

In order to find any volume quickly, you will need some order based on a logical and familiar criterion—an order in which the location of the book on the shelf indicates the style, the subject, the time period, and any other characteristic you deem significant.

A LITTLE HISTORY

Before the present Periodic Table was established, several attempts were made at classification. In the **eighteenth century**, only 31 elements were known**,** so **no classification system was needed**. Most of the elements now known were discovered in the nineteenth century. The first classification system, which we owe to the German Döbereiner, grouped the elements by threes (in **triads**) by characteristics. Examples include chlorine, bromine, and iodine; and lithium, sodium, and potassium. The atomic mass of the element in the middle was the arithmetic mean between the first and third of each triad. In 1865, the English scientist Newlands arranged the elements according to **order of atomic mass** and observed, not entirely accurately, that the **properties repeated periodically**. In 1869, the Russian Mendeleyev ordered the elements by their **atomic number** and produced the table that is still used with some small modifications.

There are presently **22 synthetic elements**; most of them are **transuranic** (with a higher atomic number than uranium).

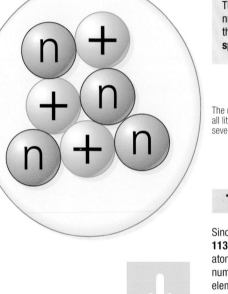

In the representative elements, the group number corresponds to the number of electrons the atom has in its last level.

ATOMIC NUMBER

The **number of protons** contained in the nucleus of an element's atom is called the atomic number. The atomic number is **specific** to each element.

The nucleus of the lithium atom has three protons, like all lithium atoms, and four neutrons. Its mass number is seven, and its atomic number is three.

THE PERIODIC TABLE

Since 1999, the Periodic Table has contained **113 elements** numbered according to their atomic number (from 1 to 114). For now number 113 does not exist, and the last four elements have no official name.

The table consists of **18 columns**. They are known as groups, and there are two types: **A (representative)** and **B (transition)**. It also consists of **seven rows** known as **periods.** One period has 2 elements, two have 8, two have 18, one has 32, and the last period has 27. The table is not finished, and it will continue to expand with time.

ATOMIC RADIUS

The radius of atoms is a **periodic property** since it varies in a regular fashion throughout periods and groups:

- As the atomic number increases throughout a period, the **atomic radius decreases.**

- As the atomic number increases throughout a group, the **atomic radius increases**.

The electroaffinity of the noble gases is zero, despite their location far to the right of the Periodic Table.

IONIZATION ENERGY

This is the **energy** that must be imparted to an atom to **take away** 1 **electron**. This energy is very high when the number of electrons the atom has in the last orbital is 8 (VIIIA, **noble gases**) and is much lower when the last orbital has only 1 electron (IA, **alkaline metals**). As the atomic number increases within a group and the atomic radius increases, the attraction the positive nucleus has for electrons in the last orbital decreases since they are farther away.

Ionization energy is also known as **ionization potential**.

ELECTROAFFINITY

When an atom **accepts an electron, it loses energy**, which is known as electroaffinity. This energy is significant in atoms located on the right of the table and **insignificant or nonexistent** in the ones that are on the left.

Another periodic property is closely related to ionization energy and electroaffinity; it is referred to as **electronegativity**. It represents the force with which an atom attracts a pair of electrons it shares with another atom.

OXIDATION NUMBER

Valence is the term used for the **combining capacity** of an element's atom. Valence has been defined many times as hydrogen atoms with which an atom reacts or may replace. The oxidation number normally coincides with the valence and tells us the **oxidation state** of an element. The oxidation number of an element that is not combined with another is always zero. If it combines and is the most electronegative one in the combination, the index is not negative but, rather, positive.

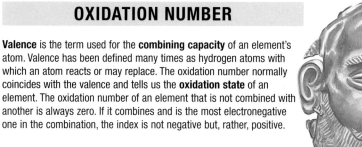

This Mycenaean funerary mask is made of gold, element 79 in the table; its symbol is Au (*aurum* in Latin).

This construction beam is made of iron, element 26 in the table; its symbol is Fe (*ferrum* in Latin).

CALCULATING THE NUMBER

A series of conditions must be kept in mind when calculating the oxidation number:

- In a compound, the sum total of the numbers of all the atoms equals zero.

- In an ion, the sum of the numbers of all the atoms equals the charge of the ion.

- Some atoms have a fixed number: H ($\pm$1), O ($-$2), elements of group IA ($+$1), of IIA ($+$2). . . .

The elements in Group IB and nitrogen are fairly inert because they have an electroaffinity less than what their position in the table would indicate and a greater ionization potential.

The Periodic Table of Elements

Groups	IA	IIA	IIIB	IVB	VB	VIB	VIIB	VIIIB			IB	IIB	IIIA	IVA	VA	VIA	VIIA	VIIIA
Periods																		
1°	G 1 **H** Hydrogen																	G 2 **He** Helium
2°	S 3 **Li** Lithium	S 4 **Be** Beryllium											S 5 **B** Boron	S 6 **C** Carbon	G 7 **N** Nitrogen	G 8 **O** Oxygen	G 9 **F** Fluorine	G 10 **Ne** Neon
3°	S 11 **Na** Sodium	S 12 **Mg** Magnesium											S 13 **Al** Aluminum	S 14 **Si** Silicon	S 15 **P** Phosphorus	S 16 **S** Sulfur	G 17 **Cl** Chlorine	G 18 **Ar** Argon
4°	S 19 **K** Potassium	S 20 **Ca** Calcium	S 21 **Sc** Scandium	S 22 **Ti** Titanium	S 23 **V** Vanadium	S 24 **Cr** Chromium	S 25 **Mn** Manganese	S 26 **Fe** Iron	S 27 **Co** Cobalt	S 28 **Ni** Nickel	S 29 **Cu** Copper	S 30 **Zn** Zinc	L 31 **Ga** Gallium	S 32 **Ge** Germanium	S 33 **As** Arsenic	S 34 **Se** Selenium	L 35 **Br** Bromine	G 36 **Kr** Krypton
5°	S 37 **Rb** Rubidium	S 38 **Sr** Strontium	S 39 **Y** Yttrium	S 40 **Zr** Zirconium	S 41 **Nb** Niobium	S 42 **Mb** Molybdenum	SIN 43 **Tc** Technetium	S 44 **Ru** Ruthenium	S 45 **Rh** Rhodium	S 46 **Pd** Palladium	S 47 **Ag** Silver	S 48 **Cd** Cadmium	S 49 **In** Indium	S 50 **Sn** Tin	S 51 **Sb** Antimony	S 52 **Te** Tellurium	S 53 **I** Iodine	G 54 **Xe** Xenon
6°	L 55 **Cs** Cesium	S 56 **Ba** Barium	S 57 **La** Lanthanum	S 72 **Hf** Hafnium	S 73 **Ta** Tantalum	S 74 **W** Tungsten	S 75 **Re** Rhenium	S 76 **Os** Osmium	S 77 **Ir** Iridium	S 78 **Pt** Platinum	S 79 **Au** Gold	L 80 **Hg** Mercury	S 81 **Tl** Thallium	S 82 **Pb** Lead	S 83 **Bi** Bismuth	S 84 **Po** Polonium	S 85 **At** Astatine	G 86 **Rn** Radon
7°	L 87 **Fr** Francium	S 88 **Ra** Radium	S 89 **Ac** Actinium	SIN 104 **Rf** Rutherfordium	SIN 105 **Db** Dubnium	SIN 106 **Sg** Seaborgium	SIN 107 **Bh** Bohrium	SIN 108 **Hs** Hassium	SIN 109 **Mt** Meitnerium	SIN 110 **Uun** Ununnilium	SIN 111 **Uuu** Unununium	SIN 112 **Uub** Ununbium	SIN 114 **Uuq** Ununcabium					

L = liquid
S = solid
G = gas
SIN = synthetic

Element 113 has not not yet been synthesized.

Lanthinides	S 58 **Ce** Cerium	S 59 **Pr** Praseodymium	S 60 **Nd** Neodymium	SIN 61 **Pm** Promethium	S 62 **Sm** Samarium	S 63 **Eu** Europium	S 64 **Gd** Gadolinium	S 65 **Tb** Terbium	S 66 **Dy** Dysprosium	S 67 **Ho** Holmium	S 68 **Er** Erbium	S 69 **Tm** Thulium	S 70 **Yb** Ytterbium	S 71 **Lu** Lutetium		
Actinides	S 90 **Th** Thorium	S 91 **Pa** Protactinium	S 92 **U** Uranium	SIN 93 **Np** Neptunium	SIN 94 **Pu** Plutonium	SIN 95 **Am** Americium	SIN 96 **Cm** Curium	SIN 97 **Bk** Berkelium	SIN 98 **Cf** Californium	SIN 99 **Es** Einsteinium	SIN 100 **Fm** Fermium	SIN 101 **Md** Mendelevium	SIN 102 **No** Nobelium	SIN 103 **Lr** Lawrencium		

Introduction

Forces and Their Effects

Motion

Energy

Heat

Fluids

Wave Motion

Sound

Optics

Electricity

Matter

Inside Matter

Mixtures

Pure Substances

Chemical Changes

Alphabetical Subject Index

BONDS

When we look at a great building, we probably do not think about the tremendous forces that the concrete has to hold up and the tensions that originate between its walls. Yet in spite of everything, the building stays standing. There are some higher forces that keep the whole building together. As long as no factor breaks this equilibrium, everything remains stable. We have seen that molecules are made up of atoms and that these atoms are united with one another even though they are separate entities. The forces that join these atoms are known as bonds.

CHEMICAL BONDS

Chemical bonds are the various types of **forces** that hold the **atoms** of a molecule **together**. There are several models for bonds; the simplest and easiest to understand is that of **valence electrons**. This model is based on the tendency of atoms to resemble the **noble gases**. Thus, according to this model, an atom will join with another atom so that the number of **electrons** in its last orbital, along with the electrons of another atom that are **shared** or **exchanged**, is the same as the ones possessed by the closest gas in the Periodic Table—usually 8.

IONIC BONDS

Let us say you go to the kitchen and find the pure white, crystalline substance that we refer to as salt. The chemical name is **sodium chloride**, and it is composed of **chlorine** and **sodium**. If we look at the table to see where these elements are located, we see that sodium (Na) is in group **IA** and that chlorine (Cl) is in **VIIA**. Thus, they have **1** and **7 electrons**, respectively, in their last orbitals. The closest noble gases to these two elements have 8 electrons in their last level, except for helium. In addition, the elements of group IA easily **lose** electrons, and the elements of VIIA easily **acquire them**. If sodium gives its only electron to chlorine, **they both have 8**, since, in the previous orbital, sodium had 8 electrons. When sodium loses 1 negative electron, it becomes **positively charged**. When the chlorine gains the electron, it becomes **negatively charged**. Both remain joined together by the force of attraction that exists between the positive and negative charges.

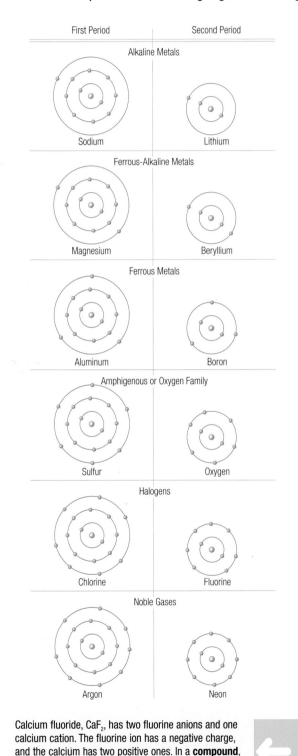

First Period — Second Period

Alkaline Metals
Sodium — Lithium

Ferrous-Alkaline Metals
Magnesium — Beryllium

Ferrous Metals
Aluminum — Boron

Amphigenous or Oxygen Family
Sulfur — Oxygen

Halogens
Chlorine — Fluorine

Noble Gases
Argon — Neon

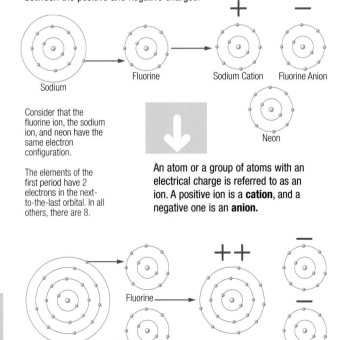

Sodium → Fluorine → Sodium Cation (+) — Fluorine Anion (−)

Neon

Consider that the fluorine ion, the sodium ion, and neon have the same electron configuration.

The elements of the first period have 2 electrons in the next-to-the-last orbital. In all others, there are 8.

An atom or a group of atoms with an electrical charge is referred to as an ion. A positive ion is a **cation**, and a negative one is an **anion**.

Fluorine (++) — (−)

Calcium

Calcium fluoride, CaF$_2$, has two fluorine anions and one calcium cation. The fluorine ion has a negative charge, and the calcium has two positive ones. In a **compound**, the **sum of the charges** always equals **zero**.

Introduction

Forces and
Their Effects

Motion

Energy

Heat

Fluids

Wave
Motion

Sound

Optics

Electricity

Matter

**Inside
Matter**

Mixtures

Pure
Substances

Chemical
Changes

Alphabetical
Subject Index

COVALENT BONDS

What do you suppose happens when neither of the two atoms loses electrons? In this case, the exchange is impossible. However, we know of substances such as chlorine (Cl_2), oxygen (O_2), and nitrogen (N_2) and many more that are **made up of 2 atoms from the right side of the Periodic Table**. In this case, we consider that in order to take on the similarity to the noble gases, they will **share** 1, 2, or even 3 or more electrons in their outer orbitals. This type of bond is referred to as covalent.

The sharing of electrons is not usually equitable in covalent bonds. The element located furthest to the right of the table retains more of the electrons and becomes a **negative pole**.

LEWIS

According to Lewis, a covalent bond is one that forms on the basis of **sharing a pair of electrons**, one from each atom, to take on the structure of a noble gas.

If the pair of shared electrons belongs to one of the atoms, the bond formed is referred to as coordinated covalent or dative bond.

- Electron
- Nucleus of Metal

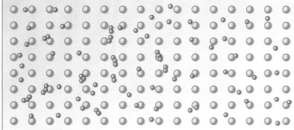

The cloud of electrons that fill up the metal is what is responsible for its most important properties.

TYPES OF COVALENT BONDS

Depending on the number of shared electrons, covalent bonds can be

- **Single**: The number of shared electrons is **a pair**, as in fluorine gas.

- **Double**: As the name suggests, the number of shared electrons is **two pairs**, as in oxygen.

- **Triple**: The number of shared electron pairs is **three pairs**, as in nitrogen.

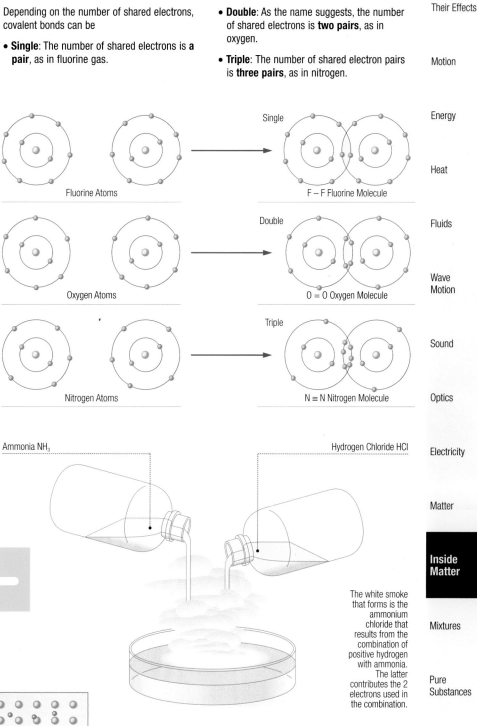

Fluorine Atoms — Single — F – F Fluorine Molecule

Oxygen Atoms — Double — O = O Oxygen Molecule

Nitrogen Atoms — Triple — N ≡ N Nitrogen Molecule

Ammonia NH$_3$

Hydrogen Chloride HCl

The white smoke that forms is the ammonium chloride that results from the combination of positive hydrogen with ammonia. The latter contributes the 2 electrons used in the combination.

METALLIC BONDS

If the joined atoms are located to the left of the table, they all have a tendency to free up electrons, and none of them can accept electrons. One of the proposed models for this type of bond is the following. All the atoms **lose** the property of their **electrons** in the outermost orbital. The electrons remain free of the nucleus. However, they become **shared**, not by the nuclei of two atoms, as in the case of covalent bonding, but by the nuclei of all the atoms that make up the solid.

GIANT STRUCTURES

When we speak of chemical bonds, we always get the impression that the atoms are bonded together in pairs, but nothing is further from the truth. Some electrons can be shared by two, three, and even millions of atoms. We have seen that the positive pole of a magnet is attracted to the negative pole. If we imagine a bag full of magnets that we have shaken up beforehand, what will the magnets be like? Stuck together in pairs or all together in a solid lump?

IONIC SOLIDS

When an atom **transfers** one of its **electrons** from its periphery to **another atom**, the latter acquires as many **negative charges** as the electrons it takes on. The atom that loses electrons, on the other hand, acquires a **positive charge**. All these ions are subject to **forces of attraction** and **repulsion** depending on whether the charges of their neighbors are the same or different. Every ion surrounds itself with as many **ions of the opposite charge** as possible based on the sizes and the proportion in which they occur.

 Metals can be considered **giant structures** since the valence electrons are shared by all the members of the crystal.

THE STRUCTURE OF SODIUM CHLORIDE

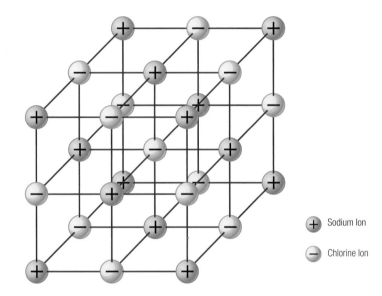

⊕ Sodium Ion

⊖ Chlorine Ion

COVALENT SOLIDS

Some authors call what we have referred to as covalent solids as **atomic solids**. This type of solid is made up of **two-** or **three-dimensional networks** in which all the atoms are joined together by **covalent bonds**. These bonds are extremely **strong**, and they give these solids some characteristics in common.

 Some chemists consider every covalent crystal to be one giant molecule.

DIAMONDS

A diamond is an **atomic solid** made of **carbon**. In this case, the carbon forms covalent bonds that are oriented toward the **vertices of an imaginary regular tetrahedron** in the center of which the carbon is located.

Diamond crystals are cut to shape by specialists.

These four bonds join the carbon with four other atoms, which join with three other atoms, and so forth until the crystal is complete.

Left: The sketch represents a carbon atom from a diamond and the directions of its bonds. *Below:* Network structure of a diamond.

Introduction

Forces and
Their Effects

Motion

Energy

Heat

Fluids

Wave
Motion

Sound

Optics

Electricity

Matter

**Inside
Matter**

Mixtures

Pure
Substances

Chemical
Changes

Alphabetical
Subject Index

NETWORK OF GRAPHITE

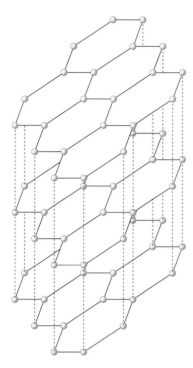

A metal is **ductile** because it can be bent, and it is **malleable** because it can be used to make sheets. Some metals are so soft that they can be cut with a pocketknife—sodium, for example.

Sodium is an alkaline metal that is very common in nature in the chloride state. Among its very numerous applications is its use as a whitener for textiles and paper (in the photo).

POLYMERS

Polymers are **organic substances** (comprised principally of **carbon** and **hydrogen**) that are formed by combining two or more similar substances known as **monomers**. There are **natural polymers**, such as cellulose, starch, glycogen, and natural rubber. Others are **synthetic**, such as plastics, synthetic rubbers, Teflon, silicone insulators, polyester, and nylon, to name just a few.

GRAPHITE

Graphite, like diamonds, is made up of **carbon**. In this case, each of the atoms is part of a **flat network** with cells in a **hexagonal** shape. These sheets are connected to one another by much weaker bonds than in diamonds, so graphite peels very readily.

Graphite is a **conductor** of electrical current; it has been used in manufacturing electrodes and pencil leads.

Silica, SiO_2, is the most abundant covalent solid in nature. **Carborundum** is one of the most common **synthetic** solids; it is similar to a diamond but with alternating atoms of carbon and silicon.

Graphite pencils
for drawing.

PROPERTIES

We cannot make sweeping generalizations about the properties of inorganic giant structures, but here is one approximation:

SOLIDS	Melting Point	Hardness	Ductile and Malleable	CONDUCTIVITY		
				In Solid State	Melted	In Solution
Ionic	Low-medium	Low	No	No	Yes	Yes
Covalent	Very high	Very high	No	No	No	No
Metallic	Medium-high	Medium-high	Yes	Yes	Yes	Yes

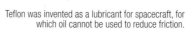

Objects that need to be very flexible are made of natural rubber, whose monomer is isoprene.

Teflon was invented as a lubricant for spacecraft, for which oil cannot be used to reduce friction.

MIXTURES

Take a look around us. Some things are made up of materials that are pure substances, but most of them are made from mixtures that may or may not be homogeneous. Bricks, floor tiles, the wood in furniture, the fabric in our clothing, and more are all mixtures. This is true not only of the things that make up our surroundings but also of the things that we drink, eat, and use. Even the water that we drink is not pure water but a solution of salts, minerals, and gases in water.

WHAT IS A SOLUTION?

A **solution** is a **homogeneous mixture** created by the **dispersion** of one or more substances known as **solutes** in another substance known as a **solvent**. If we are given a solution, we can find out which of the substances that make it up is the solvent by observing the physical states of the components. The solvent is always in the same state as the solution. In cases where two or more substances are in that state, the solvent is the one that exists in the greatest proportion of the mass.

To find out how many dissolved salts are in water, we need only look at the label on a bottle of mineral water.

If the concentration of salts inside a cell is lower than in the exterior, water leaves the cell (*left*). On the other hand, if the concentration of salts is greater, water comes into the cell (*right*).

CONCENTRATION

The concentration is the proportion between the **quantity of solute** and the **quantity of solution** or solvent.

The rubbing alcohol that we buy in drugstores and supermarkets is a solution of water in alcohol.

The concentration of saturated solutions is known as **solubility**.

TYPES OF SOLUTIONS

Solutions can be classified according to many criteria. The most common ones are the following:

- **By their concentration: diluted** if the quantity of solute is much lower than that of the solvent; **concentrated** if the quantity of solute and solvent are about equal; and **saturated** if the solvent will admit no more solute.

- **By their state of aggregation: solid, liquid, or gaseous**. We need to keep in mind that the state of a solution presumes only the state of the solvent, but the other components can be in any state.

- **By their behavior in reaction to electricity**: whether or not they are conductors. In **aqueous conducting** solutions (using water as the solvent), the solute is said to be an electrolyte and is dissociated into ions. In **nonconducting** solutions, the solute is in molecular form.

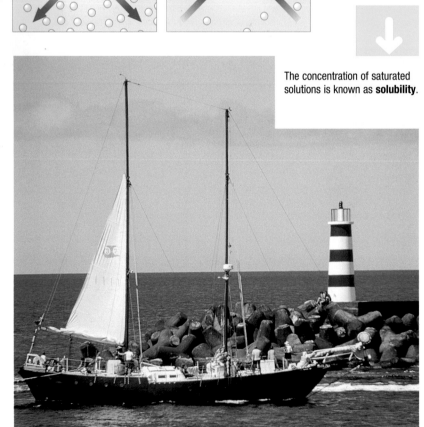

Seawater is liquid, but it contains solids such as sodium and potassium chloride, carbonates, and nitrates and also gases such as the oxygen that fish breathe.

Introduction

Forces and Their Effects

Motion

Energy

Heat

Fluids

Wave Motion

Sound

Optics

Electricity

Matter

Inside Matter

Mixtures

Pure Substances

Chemical Changes

Alphabetical Subject Index

PROPERTIES OF SOLUTIONS

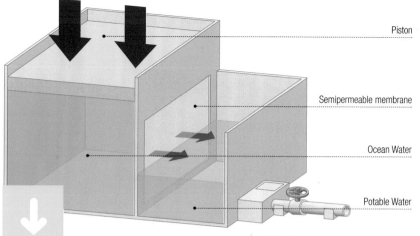

Piston

Semipermeable membrane

Ocean Water

Potable Water

Our grandmothers added bicarbonate to water when boiling chickpeas.

Nowadays potable water can be made by forcing water under pressure through a semipermeable membrane.

The properties of solutions differ from those of the pure solvent, and this difference depends in most instances on the **quantity of solute**. Let us look at a few of these properties:

- **Ebullioscopic increase**: the addition of a nonvolatile substance increases the boiling point of the solvent.

- **Cryoscopic decrease:** the addition of a nonvolatile substance decreases the solidification point of the solvent.

- **Osmotic pressure**: **osmosis** is the passage of a solvent through a semipermeable membrane. A semipermeable membrane is one that allows the solvent to pass through, but not the solute, since the pores are very small. If we have two solutions of different concentrations separated by a **semipermeable membrane**, the solvent will pass from the more dilute solution to the more concentrated one.

VARIATION IN SOLUBILITY

The solubility of solids and liquids increases when **temperature increases**, even though the increase is very different from one substance to another. Thus, when the temperature of water changes from 68° to 176°F (20° to 80°C), the quantity of salt that 100 grams of water will admit changes from 38 to 40 grams, but with sugar the quantity changes from 45 to 200 grams. Gases **reduce** their solubility as temperature rises. Solids and liquids **scarcely vary** their solubility with pressure, but the solubility of gases increases with pressure.

Water enters the cells of living beings by osmosis.

NONVOLATILE

It is said that a solute is nonvolatile when its boiling point is much higher than that of the solvent.

Salt is spread on icy roads to melt snow and ice, and antifreeze is used in the cooling water of automobiles.

Water dissolved in the air (humidity) is deposited on the coldest areas of a room and fogs up window panes.

When we remove the cap from the bottle of a carbonated beverage, the pressure decreases and liberated gas bubbles appear. If the bottle is hot, the gas escapes with greater force.

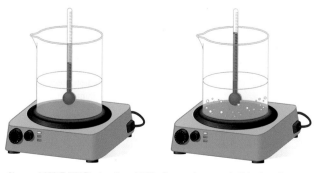

At around 122°F (50°C), when the solubility of gases decreases, bubbles from the oxygen dissolved in the water become visible.

Beer commonly has between 4.5 and 6 percent alcohol by volume.

In very specific tasks, a wide variety of units can be used, including grams per 100 cm³ of water, parts per million, and so forth.

NUMBER OF MOLES

The number of moles is calculated by dividing the mass of the solute in grams by its molecular mass.

SUSPENSIONS

If you look at river water the day after a storm, you see that it is cloudy. If we scoop some up in a glass, we see that it lacks transparency. If we let it sit, the solids gradually settle to the bottom. Thus, a suspension is the **dispersion of a solid inside a liquid**; the size of the solid particles is larger than in a solution.

After a while, cloudy water becomes transparent again because the particles in solution settle to the bottom and form a sediment.

WAYS OF EXPRESSING CONCENTRATION

In expressing concentration, any expression will do that has the quantity of solute in the numerator and the quantity of solution or solvent in the denominator. Here are the most commonly used ones:

• **Percentage of mass:**

$$\text{\% of mass} = \frac{\text{solute mass}}{\text{solution mass}} \cdot 100$$

The solution mass is the sum of the solute mass and the solvent mass.

• **Percentage by volume**: this is applicable only to gaseous solutions and partly to liquid ones; in alcoholic drinks this is written as a percentage (%):

$$\text{\% by volume} = \frac{\text{solute volume}}{\text{solution volume}} \cdot 100$$

• **Mass of solute expressed in grams per liter of solution.**

• Number of moles of solute per liter of solution. This is known as molarity (M).

• In the SI, the unit is the number of moles per cubic meter of solution, but it is very rarely used.

After a storm, the water in rivers becomes cloudy because of the particles in suspension that it carries.

GELS

A gel is a **mixture of a solid and a liquid** formed in such a way that the liquid remains trapped inside the solid and takes on a solid but flexible texture. Examples of gels include many desserts (such as custard, gelatin, pudding, and others), body washes, domestic and industrial detergents, and many sunscreen lotions.

Suspensions, like aerosols, can be transparent. However, when light passes through them, they become visible when the light is reflected by the dispersed particles.

Custard is a gel, but the cream that is often served with it is an emulsion.

Introduction

Forces and
Their Effects

Motion

Energy

Heat

Fluids

Wave
Motion

Sound

Optics

Electricity

Matter

Inside
Matter

Mixtures

Pure
Substances

Chemical
Changes

Alphabetical
Subject Index

EMULSIONS

An emulsion is a **combination of two liquids that do not mix together**, such as water and oil. Emulsions have a short life since the tiny drops of dispersed liquid gradually get back together to make up larger drops and separate out of the dispersing liquid because of their density. In order to increase the life span of an emulsion, a substance known as an emulsifier is added to keep the droplets of the dispersed substance from getting back together. **Milk** is an emulsion, which is why when milk goes bad, the fat and the proteins separate from the whey.

FOAMS

We all know foam when we see it, but describing what it is may be a little more difficult. Foam is a **gas held in by liquid** by means of **surface tension**. When we speak of foam, we automatically think of carbonated drinks such as bubbling wines, colas, soft drinks, and soapy water; but there are also natural foams or suds such as those that form in the waterfalls of rivers. There are even solid foams such as pumice stone.

Ice creams are emulsions of water with vegetable and animal fats plus sweeteners and preservatives.

AEROSOLS

An aerosol is a mixture similar to a suspension, but the dispersing agent is a **gas**. The dispersed substance can be a **solid**, as with smoke, which is an aerosol made up of carbon particles suspended in air. The dispersed substance can be a **liquid**, as with fog, which is an aerosol made up of tiny water droplets suspended in air.

Detergents increase the surface tension of water and facilitate the formation of suds.

The chimneys of large factories have electrified gratings to break up the aerosol and keep solids from getting into the air.

→ Many products (sprays, insecticides, colognes, and so on) for domestic and industrial use are known as aerosols because the substance is extracted from its container by means of a gaseous expellant that turns the substance into that form.

INORGANIC SUBSTANCES

Inorganic substances are those that make up the mineral world; they are not common in living beings. Oxygen and silicon are the most abundant elements in the mineral world. Humans have learned to extract from the earth the substances they deem important and separate them from the ones that are not worthwhile or useful when separated. They have also succeeded in making substances that do not exist in nature. Groups of substances with similar formulas and properties have been established to facilitate their study.

OXIDES

Oxides are a **combination** of an **element** and **oxygen**. Their principal formulas are of two types: A_2O_x or AO_y. A is the symbol of the element. If its subscript is 2, x is the oxidation number of A. If A has no subscript, y is half the oxidation number of A. We all are familiar with two oxides that have very different characteristics. **Iron oxide** is black, reddish, or yellowish. **Carbon dioxide**, which we exhale when breathing, is also present during combustion and in carbonated beverages.

Apples and foods in general contain inorganic substances in very small quantities.

The iron in the can of pineapple apparently rusts more readily than the aluminum of the kitchen utensil. In reality, aluminum oxidizes a lot, but the gray oxide remains attached.

 Of the **carbon dioxide** largely responsible for the greenhouse effect, 60 percent is **produced by humans**. Nature produces only 40 percent.

TYPES OF OXIDES

Oxides may be metallic or nonmetallic:

- **Metallic oxides** can form by the direct action of **oxygen** on a **metal** or by the **decomposition of salts** from that metal. Some metals oxidize violently. Sodium, for example, produces flames when it oxidizes. Others, like iron and aluminum, oxidize slowly. Some, known as noble metals, such as gold and platinum, do not form oxides with oxygen.

- **Oxides of nonmetals** can also be produced by direct action of oxygen on a **nonmetal** or by the decomposition of salts and acids.

Generally, they are **gases**, although they include silicon, which is one of the hardest solids and has a high melting point.

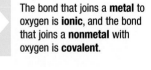 The bond that joins a **metal** to oxygen is **ionic**, and the bond that joins a **nonmetal** with oxygen is **covalent**.

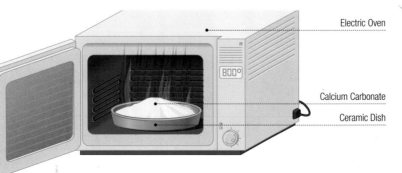

Electric Oven

Calcium Carbonate

Ceramic Dish

When calcium carbonate ($CaCO_3$) is heated over 1,472°F (800°C), it breaks down into a metallic oxide (CaO) and a nonmetallic one (CO_2, which is a gas).

Sodium hydroxide, NaOH, can be obtained from the very violent reaction of metallic sodium with water. The hydrogen that is produced burns vigorously in the air.

Introduction

Forces and
Their Effects

Motion

Energy

Heat

Fluids

Wave
Motion

Sound

Optics

Electricity

Matter

Inside
Matter

Mixtures

**Pure
Substances**

Chemical
Changes

Alphabetical
Subject Index

HYDROXIDES

Hydroxides form by the reaction of a **metallic oxide with water**. Their formula is always of the type $M(OH)_x$, where M is the symbol of the metal, and x, as previously noted, is the oxidation number of the metal. If x equals 1, it is neither indicated nor put in parentheses. Later on we will see that hydroxides are **bases**. They are designated with the name of the metal plus the word *hydroxide*, and the oxidation number of the metal is given in parentheses in Roman numerals.

THE NAMES OF OXIDES

Metallic oxides are designated using the name of the metal followed by the word *oxide*, with the oxidation number of the metal written in Roman numerals in parentheses: Fe_2O_3, iron (III) oxide; CuO, copper (II) oxide. If the metal has just one oxidation number, it is not indicated.

Oxides of nonmetals are named using a numerical prefix attached to the word *oxide* and another numerical prefix attached to the name of the nonmetal: Cl_2O_5, dichlorine pentoxide.

 A solution of the acid H_3BO_3 is what we put into our eyes to clean and disinfect them.

ACIDS

Adding **water** to some **oxides** of nonmetals produces a chemical reaction in which the product is an acid. The formula for most acids is of the type H_xAO_y. The word *acid* produces a sense of respect or even fear. The respect is justified. However, we should know that there are dangerous acids such as **sulfuric acid**

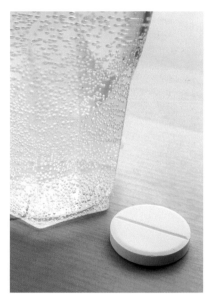

There are acids that contain no oxygen. They are known as **hydracids,** such as HCl (hydrochloric acid) and HF (hydrofluoric acid), which is the only acid that attacks glass.

(H_2SO_4) and **nitric acid** (HNO_3) and others that we calmly ingest, such as **diluted acetic acid**, which we call vinegar; citric acid, found in oranges, grapefruit, and lemons; and **acetylsalysilic acid**, which we know as aspirin.

There are several types of acid: vinegar is acetic acid, oranges and lemons contain citric acid (which is why they are known as citrus fruits), and aspirin is in reality acetylsalysilic acid.

SALTS

Salts are obtained by replacing the **hydrogens of an acid with a metal.** They can also be produced by the reaction of an acid with a hydroxide or with an oxide to produce water simultaneously. All salts are **solids** at **room temperature** and are part of most of the minerals that make up the earth.

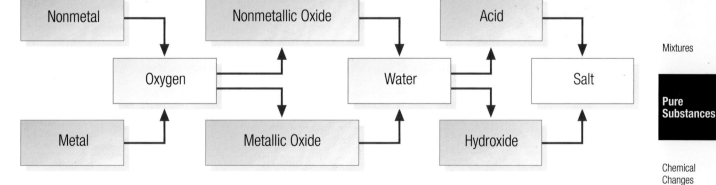

Some elements, such as aluminum, behave ambiguously. Their oxides behave like acidic or basic oxides depending on whether the substance they contact is more or less acidic than they are.

Not all oxides react with water. Some, such as NO_2 and Cr_2O_3, do not even dissolve in water.

CARBON COMPOUNDS

Carbon compounds make up such a large family within chemistry that they deserve a section specifically devoted to them. The study of carbon compounds is known as **organic chemistry** or the chemistry of life. In most cases, carbon compounds are part of living beings or we have to ascribe their origin to living beings. Even the fossil fuels such as coal and petroleum originated with living beings. Organic chemistry also includes substances based on their properties and similarities of formulas.

Petroleum (the photo shows an oil platform in the Persian Gulf) is a carbon compound and a fossil fuel.

HYDROCARBONS

The hydrocarbon function is the simplest of all. The molecules of hydrocarbons are made up exclusively of **carbon** and **hydrogen** atoms linked by covalent bonds. If all their bonds are **simple**, we say that they are **saturated**, and they make up the **alkanes.** If they contain double or triple bonds, they are **unsaturated** and make up the **alkenes** or the **alkynes**, respectively.

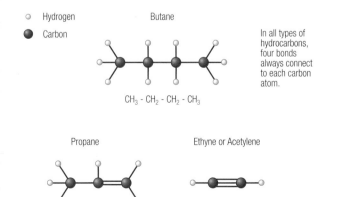

○ Hydrogen
● Carbon

Butane

In all types of hydrocarbons, four bonds always connect to each carbon atom.

$CH_3 - CH_2 - CH_2 - CH_3$

Propane

$CH_3 - CH = CH_2$

Ethyne or Acetylene

$CH \equiv CH$

CARBON CHAINS

Carbon, element number 6 on the Periodic Table, has one quality that is repeated only in boron and silicon, and then to a lesser degree. This property is that of **connecting** with other atoms of its kind and forming **linear**, **branched**, and even **cyclical** chains. The length of the chain gives the organic compound most of its physical properties: melting temperature, boiling temperature, density, and so forth.

TYPES OF CARBON CHAINS

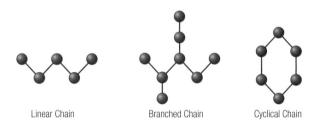

Linear Chain   Branched Chain   Cyclical Chain

FUNCTIONAL GROUPS

The **chemical properties** of organic compounds are due to a small number of atoms that constitute what is known as the functional group. If two compounds contain the same number and type of atoms, the compounds will have similar physical properties but different chemical properties if the noncarbon atoms are grouped differently. These different groups are called functional groups.

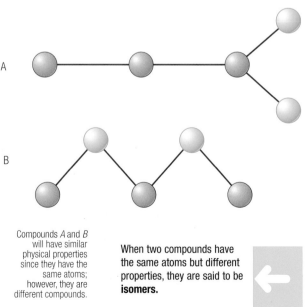

A

B

Compounds *A* and *B* will have similar physical properties since they have the same atoms; however, they are different compounds.

When two compounds have the same atoms but different properties, they are said to be **isomers.**

Introduction

Forces and
Their Effects

Motion

Energy

Heat

Fluids

Wave
Motion

Sound

Optics

Electricity

Matter

Inside
Matter

Mixtures

Pure
Substances

Chemical
Changes

Alphabetical
Subject Index

NOMENCLATURE

Hydrocarbons are named using a prefix that refers to the **length** of the chain and a suffix that depends on the **type of bond**. If there is any double or triple bond, a number will indicate it.

Number of Carbon Atoms	Prefix	Type of Hydrocarbon	Suffix	Sample Formula	Sample Name
1	Meth-	Alkane	-ane	CH_3-CH_2-CH_3	Propane
2	Eth-	Alkene	-ene	CH_2=CH-CH_2-CH_3	1-Butene
3	Prop-	Alkyne	-yne	CH_3-C≡C-CH_2-CH_2	2-Pentyne
4	But-				
5	Pent-				
…	…				

Alkenes are also known as **ethylenic** hydrocarbons, and alkynes are also called **acetylenic** hydrocarbons.

COMPOUNDS WITH OXYGEN

By means of **replacement** or **addition**, hydrocarbons give rise to a series of compounds that contain oxygen in addition to carbon and hydrogen:

- **Alcohols:** These compounds can be considered products of the replacement of a hydrogen atom in a hydrocarbon by a hydroxyl group (–OH). They are named the same way as the hydrocarbons are but the suffix of the hydrocarbon is replaced by the **suffix -ol**.

- **Aldehydes:** An aldehyde is formed when an alcohol that has the –OH at the end of the chain is oxidized; the functional group of the aldehydes is the carbonyl group (–CHO). Their name is the same as that for alcohols but with the **suffix -al**.

- **Ketones:** If the –OH group is located at a carbon atom joined to two other carbon atoms, it produces a ketone when it oxidizes. Ketones are named like the aldehydes but with the **suffix -one**; their group also is the carbonyl group.

- **Acids:** When an aldehyde oxidizes, a carboxylic acid is formed; it is given that name because the distinguishing group is the **–COOH** group, known as carboxyl.

Ethers result from combining two alcohols; **esters** are the product of a reaction between an alcohol and an acid.

Methanol or formalin is used to preserve insects and small, lower animals.

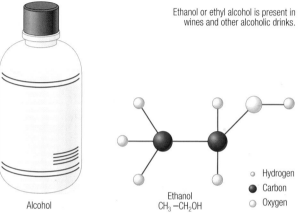
Ethanol or ethyl alcohol is present in wines and other alcoholic drinks.

Alcohol

Ethanol
CH_3 –CH_2OH

○ Hydrogen
● Carbon
○ Oxygen

NITROGEN COMPOUNDS

When **ammonia**, an inorganic compound with the formula NH_3, reacts with organic compounds, it gives rise to a series of compounds that contain nitrogen in addition to carbon and hydrogen. If one hydrogen atom of a hydrocarbon is replaced by an **amine** group, –NH_2, an **amine** is produced from the ammonia. If this same group replaces an –OH of a carboxyl group in an acid, the product is an **amide**.

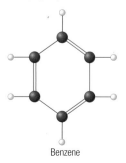

Benzene is a cyclical hydrocarbon, C_6H_6. Its derivatives are very common in nature, and they constitute what are known as the **aromatic compounds**.

Benzene

Vinegar is a solution of acetic acid; it is produced from the ethanol of wine or fruits (apples, pears, melons, and so on).

Vinegar

Acetic Acid
CH_3 –COOH

○ Hydrogen
● Carbon
○ Oxygen

THE CHEMISTRY OF LIFE

It would be no exaggeration to state that nature, both living and inert, behaves like a huge physical and chemical laboratory. Think of the motion of the continents, the evaporation of water, the rain, the transformations that take place under high pressure deep in the deep earth, and so forth.

Living beings, plants and animals, are the sum of chemical and physical phenomena: force, motion, levers, and others. At every moment in every living being, thousands of chemical reactions are going on, and chemical substances provide a living creature's structure and energy.

CARBOHYDRATES

The carbohydrates, which are also known as **glycides** because of the sweet taste some of them have, are made up of carbon, hydrogen, and oxygen. These last two are in the same proportion as in water, in other words, twice the number of hydrogen atoms than of oxygen. Among the carbohydrates are the most familiar sugars such as **glucose** and **sucrose**, which we use to sweeten foods. Carbohydrates can be **monosaccharides** or **polysaccharides**. The polysaccharides are formed from the union of two or more monosaccharides.

The potato is really the tubercle of a plant; it is very rich in starch.

If secondhand paper were recycled, the disappearance of the forests and the associated climatic changes would slow down.

Table sugar is a disaccharide produced by combining glucose and fructose; it is called sucrose.

When glucose is oxidized in the **cellular respiration** of animals, it produces the **energy** they need for their activities.

POLYSACCHARIDES

The most important polysaccharides in living beings are the following:

- **Starch:** Its giant molecule is made up of hundreds of **glucose molecules**. When an organism has too much glucose, it produces starch as a reserve. When it needs glucose, it breaks down the starch molecule by means of a hydrolysis reaction. Starch is the energy reserve of plants.

- **Glycogen:** This is very similar to starch, but it is found in animals.

- **Chitin**: This is a derivative of polysaccharides. It contains nitrogen and is the main component of the exoskeleton of arthropods such as grasshoppers.

- **Cellulose**: This also is a polysaccharide from glucose. It is distinguished from starch by the fact that its molecules are linear, long, and rigid. This is the most important component in plant cells.

CARBOHYDRATE FORMATION

The job of green plants is to form carbohydrates by the reaction known as **photosynthesis**, which is represented by the simplified equation CO_2 + H_2O + light → carbohydrate + O_2. Plants get carbon dioxide from the air, water from the soil through their roots, and light from the sun.

The cellulose extracted from the wood of trees is used in manufacturing paper.

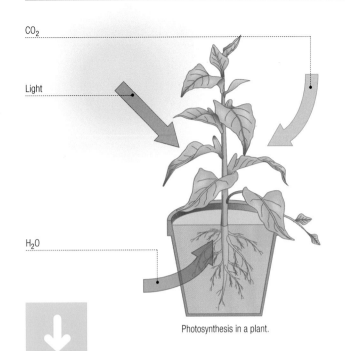

CO₂

Light

H₂O

Photosynthesis in a plant.

Petroleum spills in the ocean are largely responsible for the accumulation of CO_2 in the atmosphere since they keep the green algae from carrying out photosynthesis.

In fats and oils, the fatty acid may have all single bonds (**animal fats**) or some double or triple bonds (**vegetable fats**).

Olive oil is an edible lipid that is produced by pressing olives. Vegetable oils are unsaturated and healthier than animal fats.

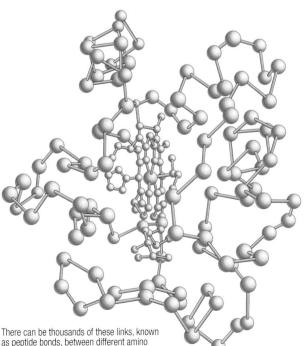

There can be thousands of these links, known as peptide bonds, between different amino acids in a protein molecule.

LIPIDS

Lipids are materials that animals and plants use for **energy reserves**; they use lipids when the supply of polysaccharides is exhausted. Lipids are **fats, oils, waxes,** and **steroids.** Fats are the ester formed in the chemical combination between a fatty acid (containing from 17 to 25 carbon atoms) and glycerin (glycerol). Fats that are in a liquid state at ambient temperature are known as oils. Waxes can be of plant or animal origin and are made up of an alcohol and a highly complex fatty acid. **Steroids** are lipids that contain four organic rings; one steroid is **cholesterol**.

PROTEINS

Proteins are formed by the union of substances known as **amino acids**. In addition to carbon, hydrogen, and oxygen, amino acids contain nitrogen and other elements such as iron, chromium, and sulfur. The properties of proteins depend on the amino acids that make them up and their relative positions. They constitute part of the **muscle fibers**, the biological catalysts (**enzymes**), the blood (**hemoglobin**), and the **genetic material** that transmits the traits from parent to child.

Proteins are not only part of our muscular and circulatory systems but also make up an important part of our genetic material, which passes on the traits of our predecessors.

Introduction

Forces and Their Effects

Motion

Energy

Heat

Fluids

Wave Motion

Sound

Optics

Electricity

Matter

Inside Matter

Mixtures

Pure Substances

Chemical Changes

Alphabetical Subject Index

THE MOST IMPORTANT CHEMICAL REACTIONS

As we have seen so far, chemists know the structure of matter and the ways to present it. In nature, matter undergoes certain changes that totally or partially modify its properties. By observing nature, humans have learned to carry out these same changes and some others that serve their needs. All chemical reactions have one thing in common: a considerable variation in energy.

MIXTURES AND CHEMICAL REACTIONS

Often we are faced with the following uncertainty: when we place two substances into contact with one another and stir them, do they simply mix or react chemically? We can resolve that doubt using three facts:

- In a **mixture**, each substance **retains its own characteristics**. In **chemical reactions**, new substances appear with **new characteristics.**

- In a **mixture**, the substances can be present in **any proportion**. In a **chemical reaction,** the proportion of each substance that reacts is always **the same**.

- In a **mixture**, the energy variation may be nonexistent or moderate. In a **chemical reaction**, energy always **varies** significantly.

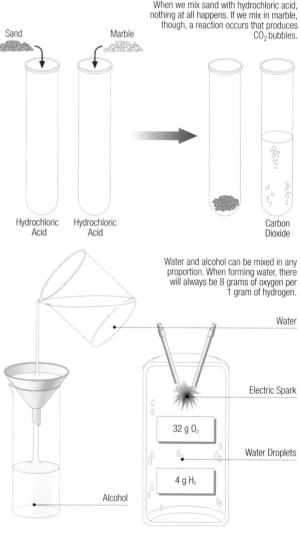

When we mix sand with hydrochloric acid, nothing at all happens. If we mix in marble, though, a reaction occurs that produces CO_2 bubbles.

Sand

Marble

Hydrochloric Acid

Hydrochloric Acid

Carbon Dioxide

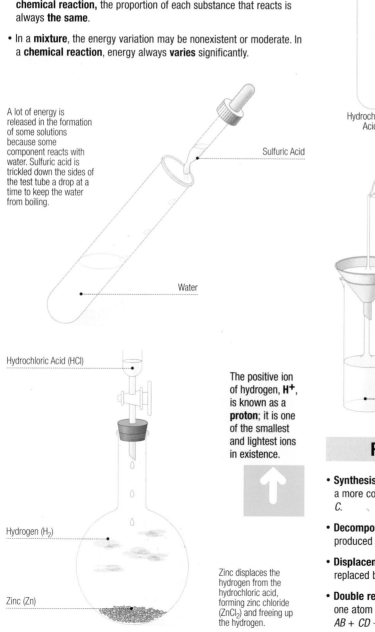

A lot of energy is released in the formation of some solutions because some component reacts with water. Sulfuric acid is trickled down the sides of the test tube a drop at a time to keep the water from boiling.

Sulfuric Acid

Water

Water and alcohol can be mixed in any proportion. When forming water, there will always be 8 grams of oxygen per 1 gram of hydrogen.

Water

Electric Spark

32 g O_2

Water Droplets

4 g H_2

Alcohol

Hydrochloric Acid (HCl)

The positive ion of hydrogen, **H⁺**, is known as a **proton**; it is one of the smallest and lightest ions in existence.

Hydrogen (H_2)

Zinc (Zn)

Zinc displaces the hydrogen from the hydrochloric acid, forming zinc chloride ($ZnCl_2$) and freeing up the hydrogen.

REACTION TYPES BY FORM

- **Synthesis**: A synthesis reaction is one where two substances produce a more complex one. These reactions can be represented as $A + B \rightarrow C$.

- **Decomposition** or **analysis:** Two or more simpler substances are produced from a single substance: $A \rightarrow B + C$.

- **Displacement**: One atom or group of atoms of a substance is replaced by another and freed up: $AB + C \rightarrow A + BC$.

- **Double replacement** or **displacement:** There is an exchange of one atom or a group of atoms between two substances: $AB + CD \rightarrow AC + BD$.

Introduction

Forces and
Their Effects

Motion

Energy

Heat

Fluids

Wave
Motion

Sound

Optics

Electricity

Matter

Inside
Matter

Mixtures

Pure
Substances

Chemical
Changes

Alphabetical
Subject Index

Burette

Hydrochloric Acid (HCl)

Sodium Hydroxide (NaOH)

Most bases are hydroxides; however, substances such as ammonium and amines are also bases.

COMBUSTION REACTION

In **classical oxidation**, the combining with oxygen is **rapid**, and **light and heat** are given off during combustion. In the combustion of hydrocarbons and oxygenated carbon compounds, **carbon dioxide** and **water** are always produced: $(CHO) + O_2 \rightarrow CO_2 + H_2O$.

In order for the balloon to rise, the air it contains must be heated; hot air weighs less than cold air.

Combustion was surely the first chemical reaction caused by humans when they discovered fire.

Even though an equilibrium reaction never ends, it appears to us that it has ended, but not all the reagents have been consumed.

EQUILIBRIUM REACTIONS

Some reactions are referred to as **complete** or irreversible. In these, some substances, called **reagents**, give rise to products. There are others in which these products can **react with one another** and form the initial substances once again. These reactions are called **reversible**. If the products **remain in contact** with one another, the reactions become **equilibrium reactions** since neither of the reactions (direct or inverse) ever ends.

ACID-BASE REACTIONS

An acid is any substance that is capable of giving up **protons** (HA). Bases are the opposites of acids since they are capable of **taking on protons** (BOH). Depending on how easily they give up or accept these protons, acids and bases can be **strong** or **weak**. When an acid reacts with a base, the result is a salt and water: $HA + BOH \rightarrow BA + H_2O$.

Hydrochloric acid drips onto sodium hydroxide and forms sodium chloride (NaCl) and water (H_2O).

OXIDATION-REDUCTION REACTIONS

Formerly it was said that an oxidation reaction occurred when there was a combination involving oxygen or a loss of hydrogen atoms. Nowadays we say that there is **oxidation** when **the oxidation number** (valence) **increases**. For that to happen, there must be a **loss of electrons**. Therefore, **reduction** occurs when **electrons are gained**. Since whenever one atom wins electrons another has to lose, these reactions are always simultaneous. These reactions are commonly written using two half-reactions:

$A + e^- \rightarrow$ products (reduction); $B \rightarrow$ products $+ e^-$ (oxidation).

Oxidation is used in jewelry making as a finish on pieces made of silver.

Acetylene (C_2H_2)

Oxygen

In an oxyacetylene torch, the acetylene burns with the oxygen and can reach a temperature sufficient to melt iron.

ENERGY IN CHEMICAL REACTIONS

If we think about it, any action that we do requires work on our part, and that involves energy consumption, and there is change in energy as a product of the action. Things happen precisely the same way in chemistry. However, many times this fact is less obvious since what we can detect is the sum of the two energies: the energy required and the energy given off. Sometimes we can see that in order to initiate a reaction, we need to apply a certain amount of energy, although later on it may come back to us multiplied many times.

WHAT ARE CHEMICAL REACTIONS LIKE?

Let us look at two cases. In the first, we scratch a match against a striker and immediately a very hot flame appears that can be used for various purposes. In the second, we wish to break down water into its two components, hydrogen and oxygen; so we circulate a direct electrical current through the water to which we have added a few drops of sulfuric acid to increase conductivity. We can observe that the two gases come bubbling up. If we cut off the current, though, no more gas is given off. The first reaction is **exothermic**, and the second one is **endothermic**.

Exothermic Reaction

Energy   Energy

Reactor

Endothermic Reaction

Energy   Energy

Reactor

Types of chemical reactions.

The combustion of a match is **exothermic** even though energy is needed to start the process (**activation energy**). If we add all the energy together, we see that more energy is given off.

WHY DOES REACTION HEAT EXIST?

We know that **substances** are made of **atoms** and that atoms are linked together by bonds. When a reaction takes place, **energy** is almost **always needed to break** the existing bonds (positive), and **energy is given off** (negative) when new bonds are formed. By adding the two energies together, we see whether there is energy left over (exothermic) or if energy was needed (endothermic). When we express the heat of a reaction, we always do it for the same quantity of matter that reacts: a **mole** of substance. If the reaction is endothermic, this heat is positive, and if it is exothermic, the heat is negative.

Water

Oxygen

Hydrogen

− +

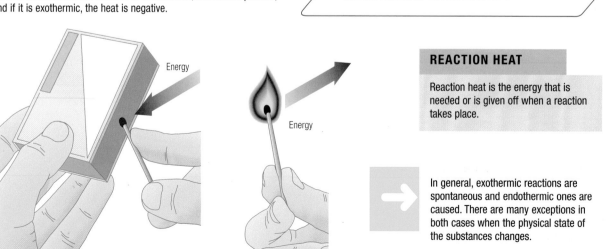

Energy

Energy

REACTION HEAT

Reaction heat is the energy that is needed or is given off when a reaction takes place.

In general, exothermic reactions are spontaneous and endothermic ones are caused. There are many exceptions in both cases when the physical state of the substances changes.

EXPANSION WORK

In order to break apart rocks and get at their secrets, miners use sticks of **dynamite.** Dynamite burns quickly and produces gases. When these gases are confined under high pressure, they expand and break apart the rock. They have performed work. This work is performed by chemical reaction, and as a result there is a decrease in the energy of the reacting substances. If there is an **increase in volume**, the **work is negative** (energy is given off), and if there is a **decrease**, the work is **positive** (energy is taken on).

Gas Outlet · Spark Plug · Gas Outlet

The cylinder of a motorcycle engine performs work thanks to the increase in volume experienced by the mixture of air and gas as it burns.

INTERNAL ENERGY

Internal energy is the **sum** of the **potential** and **kinetic energies** of all the reacting substances (all their atoms and molecules), and their variation equals the sum of the heat and the work.

ELECTRICAL ENERGY

A chemical reaction may produce electrical energy. If we mix a solution of copper (II) sulfate with zinc, several things happen: the zinc becomes coated with metallic copper, the blue color of the sulfate solution disappears, and the container heats up slightly. This is an exothermic oxidation-reduction reaction. Recall that this type of reaction can be expressed as two half-reactions. If we carry out the two reactions in different containers connected by an electrical conductor, the interchange of electrons is performed through the cable. It is an **electrolytic battery**.

In a Daniell battery, zinc (Zn) oxidizes to Zn^{2+}; in the copper sulfate, on the other hand, the copper (II) Cu^{2+} is reduced to metallic copper.

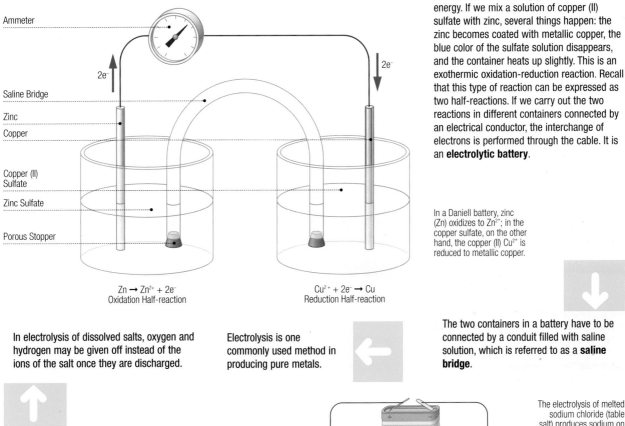

Ammeter

$2e^-$ $2e^-$

Saline Bridge

Zinc

Copper

Copper (II) Sulfate

Zinc Sulfate

Porous Stopper

$Zn \rightarrow Zn^{2+} + 2e^-$
Oxidation Half-reaction

$Cu^{2+} + 2e^- \rightarrow Cu$
Reduction Half-reaction

In electrolysis of dissolved salts, oxygen and hydrogen may be given off instead of the ions of the salt once they are discharged.

Electrolysis is one commonly used method in producing pure metals.

The two containers in a battery have to be connected by a conduit filled with saline solution, which is referred to as a **saline bridge**.

ELECTROLYSIS

Ionic substances and other dissolved or melted electrolytes conduct electrical current. If we submerge in the liquid two electrodes connected to a source of **direct current**, the positive ions go to the negative pole, which is known as the **cathode**, and the negative ions go toward the positive pole, which is referred to as the **anode**. On the electrodes, the ions lose their charge and are freed up.

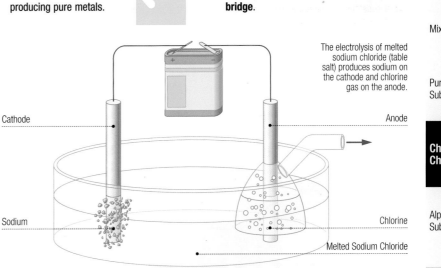

The electrolysis of melted sodium chloride (table salt) produces sodium on the cathode and chlorine gas on the anode.

Cathode

Sodium

Anode

Chlorine

Melted Sodium Chloride

Introduction

Forces and Their Effects

Motion

Energy

Heat

Fluids

Wave Motion

Sound

Optics

Electricity

Matter

Inside Matter

Mixtures

Pure Substances

Chemical Changes

Alphabetical Subject Index

CHEMICAL INDUSTRIES

The ultimate purpose of chemistry is industry. Humans investigate and observe nature, try to reproduce the observed phenomena in the laboratory, search for benefits that these phenomena may offer, study their economic return and the distribution possibilities, and finally reproduce them on a large, industrial scale. This then theoretically provides economic benefits for the owners, workers, technicians, and administrators in the industry plus benefits of another kind to the purchasers of the products.

INDUSTRY

An industry is quite a bit more than the factory where a product is **manufactured**; in every industry is a series of **departments** charged with carrying out specific tasks. All these departments have to be coordinated by a general manager in control. Here are the main departments:

- A **purchasing** department that provides the necessary raw materials.

- A **quality control** department for the raw materials.

- A technical **research and development** department—a team of scientists and engineers who design and control the manufacturing of the product plus the environmental impact, the health, and the safety of the personnel.

- A **budget and administration** department that calculates the price of the product based on production costs.

- A **manufacturing** department—a series of sections that manipulate the raw materials and make the product.

- A product **quality control** department.

- **Sales and after-sale support**.

- A **human resources** department that chooses the workers best suited to each job and takes care of training and retraining them.

In small companies, two or more departments may be combined under a single manager.

CEMENT

The quantity of cement that a country manufactures and consumes is one of the data considered in evaluating its **industrial potential**, along with its quantity of sulfuric acid and steel. The raw materials used in manufacturing cement are **calcium carbonate** (limestone), **clay**, and **sand**. They are heated and mixed in a rotating oven, and the product of this operation is called **clinker**. Once it is cooled and ground up, it can be used to produce several types of construction materials.

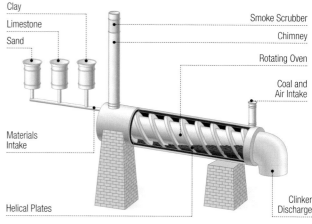

Clay
Limestone
Sand

Smoke Scrubber
Chimney
Rotating Oven
Coal and Air Intake

Materials Intake

Helical Plates

Clinker Discharge

When clinker is mixed with gypsum, it produces portland cement; when it is mixed with water and sand, it makes concrete.

The helical plates cause the burning coal to rise through the rotating oven and the raw materials to go downward through it to the product discharge point.

SODIUM HYDROXIDE

Lye is used in manufacturing soaps.

In electrolysis, sodium forms on a cathode of mercury; once it is cooled, it falls into the water to form sodium hydroxide.

Sodium hydroxide, which is popularly known as **lye**, is commonly manufactured along with chlorine since the raw materials used are sodium chloride and water. Electrolysis is used to obtain hydroxide, just as in producing sodium. Oxygen is also produced as a by-product. Lye is used in great quantities in manufacturing soaps and detergents.

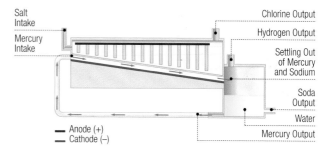

Salt Intake
Mercury Intake

Chlorine Output
Hydrogen Output
Settling Out of Mercury and Sodium
Soda Output
Water
Mercury Output

— Anode (+)
— Cathode (−)

Introduction

Forces and
Their Effects

Motion

Energy

Heat

Fluids

Wave
Motion

Sound

Optics

Electricity

Matter

Inside
Matter

Mixtures

Pure
Substances

Chemical
Changes

Alphabetical
Subject Index

AMMONIA

Ammonia is a gas that has **many applications** in the field of chemistry. It is used in the manufacture of nitric acid, ammonia salts, amines and amides, fertilizers, and dyes. It is produced from **hydrogen** and **nitrogen** from the air or by displacing the ammonium ion of ammoniac with sodium hydroxide. The **most commonly used method** is the Haber-Bosch one, which involves the inconvenience of working at very high pressure and with very pure nitrogen to avoid ruining the catalyst, which is commonly a metal such as iron or an oxide like aluminum. In spite of everything, the yield does not exceed 40 percent.

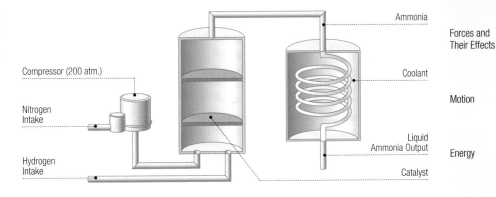

Compressor (200 atm.)

Nitrogen Intake

Hydrogen Intake

Ammonia

Coolant

Liquid Ammonia Output

Catalyst

Nitrogen oxides that have not reacted return to the raw materials intake to start the process again.

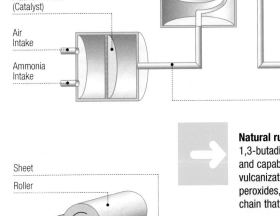

Coolant

Platinum Grate (Catalyst)

Air Intake

Ammonia Intake

Output of Nitrogen Oxides

Water

Nitric Acid Output

Nitrogen Oxides

CATALYST

A catalyst is a substance that is not consumed in a chemical reaction but does change the reaction's speed.

POLYMERS

The polymer industry has experienced more growth in the twentieth century than any other industry. The manufacturing processes and nature of polymers are extremely varied. If we try to generalize, we see three phases: production of a monomer, the condensation of monomers to make up the polymer, and treatment of the polymer to give it the shape appropriate to its use (thread, sheet, granules, and so on). The **raw materials** are commonly organic compounds from petroleum with double bonds in their molecules. Their **uses** are extremely varied: textile fibers, containers, decorations, small household utensils, automobile accessories, household appliances, and others.

Sheet

Roller

Natural rubber is a polymer of 2-methyl-1,3-butadiene (isoprene). Its chain is linear and capable of being distorted. In vulcanization reactions with sulfur or peroxides, it forms a three-dimensional chain that we call **rubber**, which is elastic.

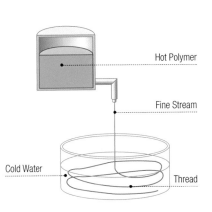

Hot Polymer

Fine Stream

Cold Water

Thread

The same polymer treated by different processing can result in totally different forms.

Many textile fibers are produced from polymers.

SECRETS FOR DISCOVERY

It can be stated confidently that humans will continue searching for the most carefully kept secrets of the universe. That is an endless pursuit, and in recent years, it has taken on a dizzying speed. People will surely improve their knowledge of nature, and technicians will search for and find applications for the new discoveries. Some of the findings that will be applied in the near future are already on the drafting tables of engineers, and the applications are taking their first steps.

NUCLEAR FUSION

Fusion energy is the **energy** given off when **two atomic nuclei join together**. For this fusion to take place, part of the mass of the particles in the nucleus disappears by becoming transformed into energy. This energy, which has a million times more power than conventional fuels, is not new. It was discovered a half-century ago. Up to now, though, it has not been applied for peaceful means since this union has been accomplished only at a temperature close to 1.8 million degrees Fahrenheit (1 million degrees Celsius). Ways of producing fusion at more reasonable temperatures are being sought.

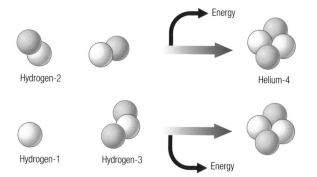

Hydrogen-2   Energy   Helium-4

Hydrogen-1   Hydrogen-3   Energy

THE MASS-ENERGY EQUIVALENCE

According to Einstein, mass can be transformed into energy: $E = \Delta m \cdot c^2$, where Δm is the loss of mass, and c is the speed of light (186,000 miles/300,000 km per second).

A digester accomplishes its mission by means of bacteria that are introduced and reproduce quickly.

In the known fusion reactions, two nuclei of hydrogen-2 combine to form helium, or one nucleus of hydrogen-1 joins with another of hydrogen-3. Hydrogen is very abundant, and helium is not a polluter.

BIOGAS

The food industry and urban areas create tons of **organic waste** every day that pollute the soil and the air. The waste can be made to decay under controlled conditions in **digesters** that produce a gas made of methane and carbon dioxide. This gas can be used as a **fuel** in electric power plants. Biogas is starting to be used under test conditions.

Selective Grate

Residue

Biogas Output

Fertilizer Output

Digester

Organic Residue

→ Biogas is a **polluter**, but the gases it produces are the same ones that would be released to the atmosphere in an uncontrolled way. The solid residue left over can be used as **fertilizer**.

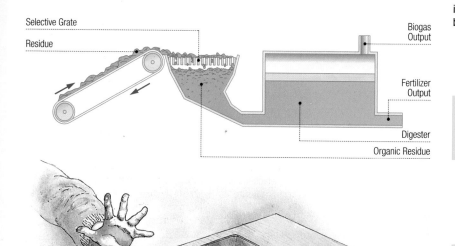

The plates of induction ranges remain cold, preventing burns or fires.

The hand pictured here would not have been burned on a magnetic induction stove since the burners remain cold.

MAGNETIC HEATING

Some modern stoves have already begun using **variable magnetic fields** to produce heat. In these stoves, steel containers are heated by a coil located underneath a **ceramic plate**. This is a very clean, precise, and safe method since only the metals are heated and the foods are cooked by the heat given off by the container.

ARTIFICIAL INTELLIGENCE

Is a modern computer intelligent? No. A computer is very adept at performing the operations that people program into it, but it has no initiative of its own. Intelligence is the characteristic of **learning** from **experience**. Currently, a few devices have some type of intelligence.

SUPERCONDUCTORS

Superconductors are substances whose electrical resistance is practically zero and that can withstand no magnetic field under a temperature known as **critical temperature.** Their full application will be realized when alloys (generally using copper) are developed with critical temperatures similar to ambient temperatures. Presently, superconductors are used in **microelectronics.**

Heat loss is a very important consideration in electronics, where current strength is very small.

Some video cameras use an image stabilizer that absorbs the involuntary motions of the photographer's hand; it works using artificial intelligence mechanisms.

SUPERFLUIDS

Superfluids are liquids that have no viscosity and no friction among their particles and are totally incapable of being compressed. So far, this behavior has been achieved only in liquid helium at −520°F (−271°C), in other words nearly at absolute zero, and also in neutron stars. Helium under these conditions will surely make it possible to study astronomical phenomena by reproducing them in a laboratory.

QUANTUM COMPUTERS

Quantum computers are based on the **tunnel effect**. This type of computer, which is much faster than conventional ones, can carry out more operations simultaneously, but it requires a **new logic**, quantum logic. Presently, this type of computer is in the early stages of experimentation, but soon it will cause a real revolution in the field of computer science.

HOLOGRAPHIC IMAGES

By means of **laser** beams and a **special emulsion**, it is possible to make a **hologram**. A hologram is no more than a three-dimensional image. A great amount of information fits into a hologram, and in the future, holograms may be used to store information. A hologram similar in size to a DVD could contain much more information than the DVD.

A hologram can contain information about the entire object, not just the front surface.

Introduction

Forces and Their Effects

Motion

Energy

Heat

Fluids

Wave Motion

Sound

Optics

Electricity

Matter

Inside Matter

Mixtures

Pure Substances

Chemical Changes

Alphabetical Subject Index

ALPHABETICAL SUBJECT INDEX

A

absolute temperatures 44
Ac actinium 71
acceleration 21
   of gravity 28
   motion 25
   unit of 21, 25
acid-base reactions 87
acids 81, 83
actinoids 71
aerosols 79
Ag silver 71
air (element) 7, 8
Al aluminum 71
alchemy 8
alcohols 83
aldehydes 83
alloys 6
Am americium 71
amino acids 85
ammonia 83, 91
amplitude 64
analysis 86
Anaximenes 7
Ar argon 71
arc, length of (unit) 25
Archimedes' principle 42–43
Aristotle 7
As arsenic 71
At astatine 71
atomic radius 70
atoms 68–69
   according to the Greeks 8
   models of 69
attraction, force of 56
Au gold 71
Avogadro's number 69

B

B boron 71
Ba barium 71
Be beryllium 71
Bh bohrium 71
Bi bismuth 71
biogas 92
Bk berkelium 71
block and tackle 17
Bohr's model 69
boiler (steam engine) 37
boiling 67
Bolos of Mendes 8
bonds 72
   chemical 72
   covalent 73
   ionic 72
   metallic 73
Boyle 9
Boyle-Mariotte's law 44
Br bromine 71
bubbles (boiling) 67

C

Ca calcium 71
calculating velocities and heights 34
calorie 39
calorific capacity 39
calorimeter 39
capillarity 41

carbohydrates 84
carbon
   chain 82
   compounds 82–83
Cartesian axes 19
C carbon 71
catalyst 91
Cd cadmium 71
Ce cerium 71
cellulose 84
cement 90
center, optical (lenses) 55
centrifuging 62
Cf californium 71
change of state 66
   laws of 67
characteristics
   of liquids 40–41
   of waves 46–47
charges, electrical 15
   types of 56
chemical bond 72
chemical reactions 86–91
chemistry of life 84–85
chitin 84
circular uniform motion 24
classification, periodic 70
Cl chlorine 71
clinker (cement production) 90
Cm curium 71
Co cobalt 71
combustion
   internal (motors) 37
   reactions 87
composition
   of forces 12
   of motions 26–27
compounds 63
   nitrogenated 83
   with oxygen 83
computer, quantum 93
concave mirror 53
concentration 76, 78
conductors 56, 58
conservation of energy 34–35
constants
   Hooke's 10
   of gravitation 14
   universal 19
contact 57
   heat propagation through 38
containers for liquids 40
contraction, relativist 19
convection (heat propagation) 38
convergent lenses 54, 55
convex mirror 53
current, electrical 58–59
   and magnetism 61
   effects of 59
   types of 58
coulomb 57
Coulomb's law 56
covalent bond 73
Cr chromium 71
crushing 62
Cs cesium 71
Cu copper 71
curves, Lissajous' 27

D

Dalton's model 68
Db dubnium 71
decanting 62
decomposition 86
   of a force 13
Democritus 8
density 42
diamond 74
diesel motor 37
difference of potential 57, 58
diffraction (sound) 49
diopters 55
direction of forces 11, 12
displacement 20, 86
   unit of 21
distance 26
distance traveled (unit of) 21
divergent lenses 54, 55
Döbereiner 70
double replacement 86
Dy dysprosium 71
dynamometer 11

E

earth (element) 7, 8
eclipse of sun and moon 51
effects of current 59
effort 30
Einstein, Albert 19, 35
electricity 56–61
electroaffinity 70
electrolysis 89
electromagnetic forces 15
electromagnetism 60–61
electromagnets 15, 61
electron 57
electronegativity 70
electrostatic
   forces 15
   phenomena 56–57
element, according to Greeks 7
elements
   representation of 63
   that characterize a motion 20–21
emulsion 79
energy 31
   conservation of 34–35
   electrical 89
   from chemical reactions 88–89
   ionization 70
   kinetic 36, 64
   mechanical 32–33
   potential 64
   types of 31
   units of 30
eolian energy 31
equilibrium 13
   reactions 87
Er erbium 71
Es einsteinium 71
Eu europium 67
evaporation 67
expansion 38, 39
   and frequency 27
   volumetric 39
extraction 62

F

fats 85
falling bodies (motion) 23
Fe iron 71
F fluorine 71
filtration 62
fission, nuclear 63
fire
   as element 7, 8
   discovery of 7
flotation 43, 62
fluids 40–45
Fm fermium 71
foam 79
focal distance (lenses) 55
focus (lenses) 55
force(s)
   applied 12
   cohesion (liquids) 40
   composition of 12
   concurrent at one point 12
   decomposition of 13
   definition of 10
   effects of 10–17
   electromagnetic 15
   electrostatic 15, 56
   magnetic 15
   measurement and moment of 11
   motion 28–29
   nuclear 15
   of friction 29
   over distance 14–15
   parallel 13
   representation of 11
   vectorial character of 11
formulas 63
fossil energy 31
Fr francium 71
frequency 47
   and expansion 27
   different 27
friction 34
   force of 29
   of rope 29
fulcrum (lever) 16
fusion 66
   nuclear 92

G

Ga gallium 71
galena 64
Galileo 9
gases 44–45, 65, 68
   ideal 65
   law of 65
Gay-Lussac's law 45
Gd gadolinium 71
Ge germanium 71
gel 78
generator, current 61
geothermal energy 31
glycides 84
glycogen 84
graphite 75
gravitation, universal law of 14
gravity 14
Greeks and matter 7, 8
grinding 62
Gutenberg, Johannes 9

H
half-light 51
heat
  absorbed by an object 39
  as energy source 36–37
  effects of 38–39
  propagation of 38
  reaction 88
  transformation into work 36
heating
  of objects 39
  magnetic 92
height 26
  calculating 34
He helium 71
Heraclitus 7
heterogeneous mixtures 62
Hf hafnium 71
Hg mercury 71
H hydrogen 71
Ho holmium 71
holographic images 93
homogenous materials 63
Hooke's law 10
Hs hassium 71
hydracids 81
hydraulic energy 31
hydrocarbons 82
hydroxide 81
  of sodium 90

I
I iodine 71
images, virtual 53
impact, elastic and inelastic 33
index of oxidation 71
induction 57
  magnetic 60
industries, chemical 90–91
In indium 71
injection (motor) 37
insulators 56
intelligence, artificial 93
intensity
  of current 59
  of field 57
  of forces 11, 12
  of sound 49
ionic bond 72
Ir iridium 71
isomers 82
isotopes 68

K
ketones 83
kinetic
  energy 32
  motion 20
  theory 64
K potassium 71
Kr krypton 71

L
La lanthanum 71
lanthanides 71
Lavoisier 9
laws
  Boyle-Mariott's 44
  Coulomb's 56

gas 45
Gay-Lussac's 45
Hooke's 10
right hand 61
universal gravitation 14
lenses 54–55
  types of 54–55
Leucippus 8
lever 16
  types of 16
light 50
  displacement of 50
  nature of 50
  speed of 19, 35
Li lithium 71
line of motion 11, 12
lipids 85
liquids 65
  characteristics of 40–41
  definition 40
Lr lawrencium 71
Lu lutecium 71
lye 90

M
machines
  simple 16–17
  thermal 37
magnetic field 15, 61
  intensity of 57
magnetic forces 15
magnetism 60–61
  and current 61
magnetite 60
magnets 60
  loading 61
magnitudes, electrical 59
mass
  atomic 69
  molecular 69
materials, homogenous 63
matter 62–75
  classification of 62–63
Md mendelevium 71
measurement of forces 11
measuring devices 61
mechanical energy 32–33
mendeleyev 70
metallic bond 73
metallurgy 6
metals, discovery of 6
Mg magnesium 71
mirrors 52–53
  types of 52
mixtures 76–79, 86
  heterogeneous 62
Mn manganese 71
models of atoms 68–69
mole 69
moment of force 11
Mo molybdenum 71
monosaccharides 84
motion 18–30
  amplitude of 26, 27
  characteristics of 20
  force and 28–29
  oscillatory 46
  perpendicular vibratory 26

relativity of 18
  types of 22–25
motor, internal combustion 37
Mt meitnerium 71

N
Na sodium 71
Nb niobium 71
Nd neodymium 71
Ne neon 71
neodymium 60
Newton (unit) 28
Newton, Isaac 9, 28
Ni nickel 71
N nitrogen 71
No nobelium 71
Np neptunium 71
nuclear forces 15
number
  atomic 70
  Avogadro's 69

O
Oersted, Christian 61
Oersted's experiment 61
oils 85
O oxygen 71
optical axis (lenses) 55
optics 50–55
Os osmium 71
oxidation number 71
oxidation-reduction reactions 87
oxides 80
  metallic 80
  nonmetallic 80
oxygen, compounds with 83

P
pair of forces 13
Pa protactinium 71
parallel (electrical current) 59
Pascal's theorem 42–43
Pb lead 71
Pd palladium 71
pendular motion 25
penumbra 51
period 23, 47
periodic motions 22–25
Periodic Table 70–71
plane, incline 17
Pm promethium 71
point
  of application (forces) 12
  of reference (motion) 19
  of support (lever) 16
poles of magnet 60
polymers 75, 91
polysaccharides 84
Po polonium 71
position (motion) 20
potential, electrical 57
potential, gravitational (energy) 32
power 16, 30
P phosphorous 71
pressure 42
  atmospheric 45
  steam 66
pressure-volume relationship 44

principles
  Archimedes' 42–43
  Newton's 28
  of relativity 19
proteins 85
Pr praesodymium 71
Pt platinum 71
pulleys 17
Pu plutonium 71
Pythagoras 7

Q
quantity of motion 33

R
radian 24
radiation (heat propagation) 38
Ra radium 71
Rb rubidium 71
Rd rhodium 71
reactions
  acid-base 87
  chemical 87
  combustion 87
  equilibrium 87
  oxidation-reduction 87
rectilinear motion 22–23
reflection
  laws of 52
  types of 52
reflection (sound) 49
refraction 54
refraction (sound) 49
relativistic contraction 19
relativity of motion 18
relays 61
renewable energy 31
repulsion, force of 56
Re rhenium 71
reference systems 18, 19
resistance
  of current 59
  of lever 16
reverberation 49
Rf rutherfordium 71
right hand, law of 61
Rn radon 71
rope friction 29
rubber, natural 91
Ru ruthenium 71
Rutherford's model 69

S
salts 81
Sb antimony 71
Schrödiger's model 69
screw 17
Sc scandium 71
secrets for discovery 92–93
series (electrical current) 59
Se selenium 71
Sg seaborgium 71
shadow 51
Si silicon 71
Sm samarium 71
Sn tin 71
solar energy 31
solenoid 61

solids 64–65
  covalent 74
  ionic 74
solubility, variation in 77
solutions 63, 76
  properties 77
  types 76
sound 48–49
  intensity of 49
  properties of 49
  speed of 48
spiral 61
Sr strontium 71
S sulfur 71
starch 84
state, change of 66
steam
  engines 37
  pressure 66
Steroids 85
structures, giant 74–75
sublimation 66
substances
  inorganic 80
  pure 63
  simple 63
superconductors 93
superfluids 93

suspensions 78
symbols 63
synthesis 86
systems of reference 18, 19

T
Tb terbium 71
Tc technetium 71
temperature and heat 38–39
tension
  on ropes 29
  surface 41
Te tellurium 71
Thales of Miletus 7
theorems
  Archimedes' 42–43
  Pascal's 42–43
theory, kinetic 64
thermal machines 37
Thompson's model 69
Th thorium 71
timbre 49
Ti titanium 71
Tl thallium 71
Tm thulium 71
tone 49
trajectory 20
  horizontal and oblique 26

motion 20, 22
transformation
  of heat into work 36
  of work into heat 36
transmutation 8
types of motion 22–23

U
U uranium 71
units
  of force 11
  of gravity 14
  of motion 23

V
vapor pressure 66
velocity
  calculating 34
  linear (unit) 25
  median 21
  motion 22
  of light 19, 35
  unit of 21
vessels, communicating 43
vibration 64
vibratory motion 23
viscosity 41
V vanadium 71

W
water (element) 7, 8
waves
  amplitude 47
  characteristics of 46–47
  length of 47
  properties of 47
  types of 46
waxes 85
weight 14
winch 17
work 30
  as modifier of energy 34
  of expansion 89
  transformation into heat 36
W tungsten 7

X
Xe xenon 71

Y
Yb ytterbium 71
Y yttrium 71

Z
Zn zinc 71
Zr zirconium 71